Algebra Essentials

Algebra Essentials

Donald Miles

MURPHY & MOORE
www.murphy-moorepublishing.com

Published by Murphy & Moore Publishing,
1 Rockefeller Plaza,
New York City, NY 10020, USA

ISBN: 978-1-63987-037-0

Cataloging-in-Publication Data

Algebra essentials / Donald Miles.
p. cm.
Includes bibliographical references and index.
ISBN 978-1-63987-037-0
1. Algebra. 2. Mathematical analysis. 3. Mathematics. I. Miles, Donald.
QA155 .A44 2022
512--dc23

For information on all Murphy & Moore Publications
visit our website at www.murphy-moorepublishing.com

MURPHY & MOORE

Contents

Preface

Algebra is the branch of mathematics which is concerned with the study of mathematical symbols and rules governing the use of these symbols. It focuses on equation solving along with abstract concepts such as fields, groups and rings. Algebra is divided into two branches, elementary algebra and advanced algebra. Elementary algebra which includes linear equations, quadratic equations and use of real or complex variables forms the basic part of the subject and is generally considered essential for study of science, engineering, medicine and economics. Advanced mathematics is covered under abstract algebra and includes concepts such as sets, identity elements and inverse elements. It is pursued only by mathematicians. The topics included in this book on algebra are of utmost significance and bound to provide incredible insights to readers. It unfolds the innovative aspects of algebra which will be crucial for the holistic understanding of the subject matter. Coherent flow of topics, student-friendly language and extensive use of examples make this book an invaluable source of knowledge.

A detailed account of the significant topics covered in this book is provided below:

Chapter 1- Algebra is the branch of mathematics which deals with mathematical symbols and the rules for manipulating these symbols. Pre-algebra, Elementary algebra, Abstract algebra, Linear algebra and Universal algebra are the various branches of algebra. This chapter serves as an introduction to algebra and its relevant aspects.

Chapter 2- The study of algebraic structures is known as abstract algebra. Algebraic structures include rings, groups, modules, fields, vector spaces and lattices. Calculations that combine two elements are called binary operations. This chapter has been carefully written to provide an easy understanding of the varied facets of abstract algebra.

Chapter 3- The branch of algebra concerning with linear equations and linear maps is called linear algebra. Linear Algebra is used for representing geometry and basic objects. The topics elaborated in this chapter will help in gaining a better perspective about linear algebra.

Chapter 4- Algebraic equations are the equations in which two expressions are set equal to each other. Polynomials are the expressions that contain variables and coefficients and involve the operations of multiplication, addition, subtraction and positive integer exponentiation of variables. This chapter closely examines the key concepts of algebraic equations to provide an extensive understanding of the subject.

Chapter 5- Geometric Algebra deals with the vector space with quadratic form and geometric product. Algebra elements called multivectors, contains scalars and vectors. Geometric Algebra is a vast subject that branches out into significant sub-disciplines. Topics such as Algebraic variety, Hilbert's Nullstellensatz and Bézout's theorem have been thoroughly discussed in this chapter.

Chapter 6- Fundamental theorems of algebra, Abel-Ruffini theorem, Binomial theorem, Chevalley-Warning theorem, Boolean Prime Ideal theorem, Rational root theorem, factor theorem are the various theorems in algebra. This chapter discusses in detail the theorems related to algebra.

It gives me an immense pleasure to thank our entire team for their efforts. Finally in the end, I would like to thank my family and colleagues who have been a great source of inspiration and support.

Donald Miles

1

Algebra: An Introduction

Algebra is the branch of mathematics which deals with mathematical symbols and the rules for manipulating these symbols. Pre-algebra, Elementary algebra, Abstract algebra, Linear algebra and Universal algebra are the various branches of algebra. This chapter serves as an introduction to algebra and its relevant aspects.

Algebra is a branch of mathematics that deals with symbols and following rules that helped in maneuvering those symbols. In elementary algebra, these symbols represent quantities without a fixed value more commonly known as variables. Simply putting it, algebra is all about finding the undefined variables into equations and finally solving them. It is that branch of mathematics that substitutes letters for numbers, and it is an algebraic equation that represents a scale on both ends on what is done. Algebra constitutes complex numbers, real numbers, matrices, vectors, and various other forms of mathematical representations.

The roots of algebra are as deep as in the 539BC. It is said that people of Mesopotamia used the then called Babylonians mathematics which took a form of advanced arithmetical system, now commonly referred to as Algebra. The Babylonians used algebra in those days, developed a course of formulas using Algebra to solve problems similar to today by using linear equations, quadratic equations, and many more similar formulas.

Eventually, with passing time, it was the Greeks who officially first raised the curtains to introduce Algebra in the third century following the footsteps of the Babylonians. Diophantus was eventually named as the Father of Algebra. However, it was Rene Descartes who gifted us what we learn and use today in the form of Algebra through his book entitled La Geometry. Now, that we gave a good dose of history and evolution of Algebra, let us talk about the importance of Algebra in real life. Yes, this is one such part of our mathematics curriculum which will and is helping many in their real lives.

General Equations of Algebra:

$$a\hat{}2-b\hat{}2=(a-b)*(a+b)$$

$$(a+b)\hat{}2=a\hat{}2+2ab+b\hat{}2$$

$$a\hat{}2+b\hat{}2=(a-b)2+2ab$$

The Importance of Algebra in Real Life

Whether you believe it or not, algebra is needed in many spheres of life. The numbers and equations gifted by Algebra are frequently used in almost everywhere in this world. Take, for example; you are out for grocery shopping. Now, the addition and subtraction are the only sources to help you keep yourself budgeted with the items in the cart. However, for those who are not convinced of what to leave and what to take a cashier can help them using some formulation of algebra.

Not many of you are aware, even the economists of every country take the help of Algebra to understand the economy better. Money is always a part that lands a person or an economy in a dilemma. In that case, algebra is the only source of help on can get through to solve the problems with how to subtract debts or loans that one has gained throughout the years.

Algebra is also considered as a 'trump card' for success in any business. Let us take an example; a person runs an online toy store, algebra is what will come in handy for him to calculate his profit margin on any given item and make up his mind whether that item is worth stocking. Algebra would also help him in calculating the lowest price he can sell an item and still make a glaring profit from it.

Also, for every student who wishes to master the statistics and calculus, algebra is a must-have thing for them. The concepts of algebra when combined with statistics and calculus are well-used in solving complex problems in one's job. Here are some of the many instances where algebra comes in handy in real life. There are various types of algebra that one should know before applying any one of it in their real life, solving problems.

Types of Algebra

There are three main types of algebraic expressions, by the position of variables, the types of functions and operations, and the behavior of their graphs. The types are:

- Monomial Expression: The algebraic expression consisting of just ONE term is known as Monomial expressions. An example of a monomial expression is 3 x 4, 3xy.
- Binomial Expression: The algebraic equation consisting of two unlikely terms are considered as Binomial expression. Examples of Binomial expressions are 5xy + 8, xyz + x3, and many more such kinds.
- Polynomial Expression: A polynomial expression is considered the ones which consist of more than one term with non-negative integral exponents of a variable. Examples of polynomial expressions are ab + bc + ca, etc.

Learning algebra at school and mastering this broad-spectrum, versatile and resourceful branch of mathematics can be a fulfilling experience for students. If students spend just a couple of hours on a regular basis, they'll pick up the basics of solving a wide range of algebraic problems. Once they imbibe the fundamentals, they can regularly practice solving elementary, advanced, abstract, commutative, and linear algebraic sums. Understanding the essentials of algebra not only enable you to complete an academic program or course effectively but equip you with wisdom and skills that you can use a lifetime.

The Versatility of Algebra Comes in Handy in Every Sphere of Life

It is not only mathematicians but even most academicians, educationists, scholars, and professionals from all other walks of life unanimously agree on the versatility of algebra. You can liken algebra to an all-purpose tool or a magic wand that can help deal with everyday problems of life.

Algebra can help you to measure your living room dimensions and determine the size of an electronic item like TV or refrigerator that'll fit comfortably in the space. Algebraic notations and symbols are crucial for deciphering and understanding musical notes, forces of gravity, behavior of solids, liquids, and gases. The myriad branches of science that have helped us comprehend and appreciate everyday natural phenomenon, use algebra extensively.

While you visit the market for buying essentials, reining in your expenses is your topmost priority. And this is where algebra can come to your aid; you assign an alphabetical or numeric symbol to each expense head (item), and quickly calculate the distinct value of each, given your total expenditure. If you've chosen to pursue a study program in high school, college, and university where mathematics is a compulsory subject, you'll indispensably need algebra.

Algebra can Boost and Bolster Logical Reasoning

You can bet that you'll be able to think more logically and with greater clarity if and when you practice doing algebraic sums habitually. To be specific, try to solve at least 5-10 problems relating to different sub-branches of algebra such as elementary algebra, linear algebra, abstract algebra, and Boolean algebra, regularly. You'll discover to your amazement that you're able to solve problems pertaining to other branches of mathematics with remarkable ease. Algebra could go a long way in enabling you to fine-tune and sharpen your cognitive skills and reasoning capability, eventually letting you complete your study program successfully.

Students can use Algebra for Understanding other Mathematical Branches

Students looking to make a career in the media, social science, actuarial science, health and wellness, engineering, medicine, or R & D will need to master other branches of mathematics, especially calculus and statistics. Having a robust understanding of algebra enables students to pick up the basics of statistics, calculus, and other mathematical branches.

High School Algebra: Everyday Applications

It is an open secret that the majority of school students dread mathematics chiefly because solving mathematical problems poses a near insurmountable challenge. Most of the students are unwilling or do not have the tenacity to exercise their intellect for working out the problems. However, algebra is one broad part or discipline of mathematics that many students find relatively easier than other mathematical branches like calculus, trigonometry or logarithms.

No wonder, a good number of secondary and high school students find algebra remarkably stimulating. The parents and guardians of school students might recollect taking a more than usual interest in grappling with algebraic basics like quadratic equations, linear equations, and so on. Students will surely agree that the excitement associated with studying mathematical notations and symbols and maneuvering these representations to solve algebraic problems is incomparable.

To put things in perspective, algebra not only plays a key role in a student's academic life but also finds across-the-board applications in all other spheres of life. You'd invariably need algebra for studying specific subjects in high school and college as well as during your undergraduate and postgraduate years. Additionally, you'll have to excel in this branch of mathematics for doing well in competitive exams which in turn will enable you to register for a professional program.

Moreover, you'll need to solve algebraic sums, problems, and equations ingeniously and quickly for qualifying in job-oriented competitive and entrance exams. Then again, you'll find algebra indispensable for estimating the price of various essential goods and services and keeping track of your expenses. In short, algebra as an integral part of mathematics touches almost every aspect of our life.

Indispensable for Pursuing Higher Education

It does not bear emphasis algebra will be with you all the way throughout your academic period, especially if you opt for science or commerce stream. For instance, if you choose to study pure science subjects like physics, chemistry, and mathematics in high school, college, and university, you should be prepared to tackle advanced algebra. You'll also have to deal with algebra if economics, statistics, music, psychology, and numerous other subjects are included in your study programs.

Are you planning to work after clearing 10+2 in order to fund your undergraduate and/or postgraduate studies? Yes, you'll find enough jobs that will easily enable you to finance your higher education as well as leave you with a substantial amount for building a nest egg. Though you may not need to be a graduate or postgraduate for applying to these job positions, you'll need to have a high school certificate and be well-versed in algebra.

Essential for Staying in Business

While the majority of students go on to choose a profession of their liking, a good number of students opt to set up a startup. Regardless of the nature of your business, you'll have to maintain a daily record of your earnings and expenditures so as to determine if you're making a profit or incurring losses. Here again, algebra will help you work out the profit margin or extent of losses, and let you keep track of your revenues.

Applications in Everyday Life

Learning algebra in the early years, especially when you're in secondary or high school serves you well throughout your life. Apart from helping you to pass specific subjects during your academic life, algebra plays a vital role in letting you breeze through competitive exams, written tests and interviews for a wide range of job positions. Algebra comes to your aid when you decide to settle down; helps you to choose a home as per your means, aids in preparing a monthly expenditure budget, facilitates choosing the best family insurance plan, and so on.

Branches of Algebra

Algebra is divided into different sub-branches such as elementary algebra, advanced algebra, abstract algebra, linear algebra, and commutative algebra.

Algebra 1 or Elementary Algebra

Elementary Algebra covers the traditional topics studied in a modern elementary algebra course.

Arithmetic includes numbers along with mathematical operations like +, -, x, ÷. But in the field of algebra, the numbers are often represented by the symbols and are called variables such as x, a, n, y. It also allows the common formulation of the laws of arithmetic such as, a + b = b + a and it is the first step that shows the systematic exploration of all the properties of a system of real numbers.

The concepts coming under the elementary algebra includes variables, evaluating expressions and equations, properties of equalities and inequalities, solving the algebraic equations and linear equations having one or two variables, and so on.

Algebra 2 or Advanced Algebra

This is the intermediate level Algebra. This algebra has a high level of equations to solve as compared to pre-algebra. Advanced algebra will help you to go through the other parts of algebra such as:

- Equations with inequalities.
- Matrices.
- Solving system of linear equations.
- Graphing of functions and linear equations.
- Conic sections.
- Polynomial Equation.
- Quadratic Functions with inequalities.
- Polynomials and expressions with radicals.
- Sequences and series.
- Rational expressions.
- Trigonometry.
- Discrete mathematics and probability.

Abstract Algebra

Abstract algebra is one of the divisions in algebra which discovers the truths relating to algebraic systems independent of specific nature of some operations. These operations, in specific cases, have certain properties. Thus we can conclude some consequences of such properties. Hence this branch of mathematics called abstract algebra. Abstract algebra deals with algebraic structures like the fields, groups, modules, rings, lattices, vector spaces, etc. The concepts of the abstract algebra are below:

- Sets: Sets is defined as the collection of the objects that are determined by some specific property for a set. For example: A set of all the 2 × 2 matrices, the set of two-dimensional vectors present in the plane and different form of finite groups.

- Binary Operations: When the concept of addition is conceptualized, it gives the binary operations. The concept of all the binary operations will be meaningless without a set.
- Identity Element: The numbers 0 and 1 are conceptualized to give the idea of an identity element for a specific operation. Here, 0 is called the identity element for the addition operation, whereas 1 is called the identity element for the multiplication operation.
- Inverse Elements: The idea of Inverse elements comes up with a negative number. For addition, we write "-a" as the inverse of "a" and for the multiplication, the inverse form is written as "a^{-1}".
- Associativity: When integers are added, there is a property known as associativity in which the grouping up of numbers added does not affect the sum. Consider an example, $(3 + 2) + 4 = 3 + (2 + 4)$.

Linear Algebra

Linear algebra is a branch of algebra which applies to both applied as well as pure mathematics. It deals with the linear mappings between the vector spaces. It also deals with the study of planes and lines. It is the study of linear sets of equations with the transformation properties. It is almost used in all the areas of Mathematics. It concerns the linear equations for the linear functions with their representation in vector spaces and through the matrices. The important topics covered in linear algebra are as follows:

- Linear equations.
- Vector Spaces.
- Relations.
- Matrices and matrix decomposition.
- Relations and Computations.

Commutative Algebra

Commutative algebra is one of the branches of algebra that studies the commutative rings and its ideals. The algebraic number theory, as well as the algebraic geometry, depends on the commutative algebra. It includes rings of algebraic integers, polynomial rings, and so on. There are many other areas of mathematics that draw upon commutative algebra in different ways such as differential topology, invariant theory, order theory, and general topology. It has occupied a remarkable role in modern pure mathematics.

Algebraic Structure

An algebraic structure is any set upon which "arithmetic-like" operations have been defined. The importance of such structures in abstract mathematics cannot be overstated. By recognized a given set S as an instance of a well-known algebraic structure, every result that is known about that abstract algebraic structure is then automatically also known to hold for S. This utility is, in large part, the main motivation behind abstraction.

Binary Operations and Scaling Operations

When discussing an arbitrary nonempty set S, you should never assume that S has any type of "structure" (algebraic or otherwise) unless the context suggests differently. Put another way, the elements in S can only every really be related to each other in a subjective manner. E.g., if we take S = {Alice, Bob, Carol}, then there is nothing intrinsic in the definition of S that suggests how these names should objectively be related to one another.

If, on the other hand, we take $S = \mathbb{R}$, then you have no doubt been conditioned to expect that a great deal of "structure" already exists within S. E.g., given any two real numbers $r_1, r_2 \in \mathbb{R}$, one can form the sum $r_1 + r_2$, the difference $r_1 - r_2$, the product r_1r_2, the quotient r_1/r_2 (assuming $r_2 \neq 0$), the maximum $\max\{r_1, r_2\}$, the minimum $\min\{r_1, r_2\}$, the average $(r_1 + r_2)/2$, and so on. Each of these operations follows the same pattern: take two real numbers and "combine" (or "compare") them in order to form a new real number. Moreover, each of these operations imposes a sense of "structure" within R by relating real numbers to each other. We can abstract this to an arbitrary nonempty set as follows:

Definition: A binary operation on a nonempty set S is any function that has as its domain S × S and as its codomain S. In other words, a binary operation on S is any rule $f: S \times S \to S$ that assigns exactly one element $f(s_1, s_2) \in S$ to each pair of elements $s_1, s_2 \in S$. We illustrate this definition in the following examples:

Addition, subtraction, and multiplication are all examples of familiar binary operations on R. Formally, one would denote these by something like,

$$+: \mathbb{R} \times \mathbb{R} \to \mathbb{R},\ -: \mathbb{R} \times \mathbb{R} \to \mathbb{R}, \text{ and } *: \mathbb{R} \times \mathbb{R} \to \mathbb{R}, \text{ respectively.}$$

Then, given two real numbers $r_1, r_2 \in \mathbb{R}$, we would denote their sum by $+(r_1, r_2)$, their difference by $-(r_1, r_2)$, and their product by $*(r_1, r_2)$. (E.g., $+(17, 32) = 49$, $-(17, 32) = -15$, and $*(17, 32) = 544$). However, this level of notational formality can be rather inconvenient, and so we often resort to writing $+(r_1, r_2)$ as the more familiar expression $r_1 + r_2$, $-(r_1, r_2)$ as $r_1 - r_2$, and $*(r_1, r_2)$ as either $r_1 * r_2$ or r_1r_2.

The division function $\div: \mathbb{R} \times (\mathbb{R} \setminus \{0\}) \to \mathbb{R}$ is not a binary operation on $\mathbb{R}$ since it does not have the proper domain. However, division is a binary operation on $\mathbb{R} \setminus \{0\}$.

Other binary operations on $\mathbb{R}$ include the maximum function $\max: \mathbb{R} \times \mathbb{R} \to \mathbb{R}$, the minimum function $\min: \mathbb{R} \times \mathbb{R} \to \mathbb{R}$, and the average function $(\cdot + \cdot)/2: \mathbb{R} \times \mathbb{R} \to \mathbb{R}$.

An example of a binary operation f on the set S = {Alice, Bob, Carol} is given by,

$$f(s_1, s_2) = \begin{cases} s_1 & \text{if } s_1 \text{ alphabetically precedes } s_2, \\ \text{Bob} & \text{otherwise.} \end{cases}$$

This is because the only requirement for a binary operation is that exactly one element of S is assigned to every ordered pair of elements $(s_1, s_2) \in S \times S$.

Even though one could define any number of binary operations upon a given nonempty set, we are generally only interested in operations that satisfy additional "arithmetic-like" conditions. In other words, the most interesting binary operations are those that, in some sense, abstract the salient properties of common binary operations like addition and multiplication on $\mathbb{R}$. At the same time, though, binary operations can only be used to impose "structure" within a set. In many settings, it is equally useful to additional impose "structure" upon a set. Specifically, one can define relationships between elements in an arbitrary set as follows:

Definition: A scaling operation (a.k.a. external binary operation) on a nonempty set S is any function that has as its domain $\mathbb{F} \times S$ and as its codomain S, where $\mathbb{F}$ denotes an arbitrary field. (As usual, you should just think of $\mathbb{F}$ as being either $\mathbb{R}$ or $\mathbb{C}$).

In other words, a scaling operation on S is any rule $f : \mathbb{F} \times S \to S$ that assigns exactly one element $f(\alpha, s) \in S$ to each pair of elements $\alpha \in \mathbb{F}$ and $s \in S$. This abstracts the concept of "scaling" an object in S without changing what "type" of object it already is. As such, $f(\alpha, s)$ is often written simply as αs. We illustrate this definition in the following examples:

Scalar multiplication of n-tuples in $\mathbb{R}^n$ is probably the most familiar scaling operation to you. Formally, scalar multiplication on $\mathbb{R}^n$ is defined as the following function:

$$(\alpha,(x_1,\ldots,x_n)) \mapsto \alpha(x_1,\ldots,x_n) = (\alpha x_1,\ldots,\alpha x_n), \forall \alpha \in \mathbb{R}, \forall (x_1,\ldots,x_n) \in \mathbb{R}^n.$$

In other words, given any $\alpha \in \mathbb{R}$ and any n -tuple $(x_1,\ldots,x_n) \in R^n$, their scalar multiplication results in a new n-tuple denoted by $\alpha(x_1,\ldots,x_n)$. This new n-tuple is virtually identical to the original, each component having just been "rescaled" by α.

Scalar multiplication of continuous functions is another familiar scaling operation. Given any real number $\alpha \in \mathbb{R}$ and any function $f \in C(\mathbb{R})$, their scalar multiplication results in a new function that is denoted by αf, where αf is defined by the rule,

$$(\alpha f)(r) = \alpha(f(r)), \forall r \in \mathbb{R}.$$

In other words, this new continuous function $\alpha f \in C(\mathbb{R})$ is virtually identical to the original function f; it just ``rescales" the image of each $r \in \mathbb{R}$ under f by α.

The division function $\div : \mathbb{R} \times (\mathbb{R} \setminus \{0\}) \to \mathbb{R}$ is a scaling operation on $\mathbb{R}\setminus\{0\}$. In particular, given two real number $r_1, r_2 \in \mathbb{R}$ and any non-zero real number $s \in \mathbb{R}\setminus\{0\}$, we have that $\div(r_1, s) = r_1(1/s)$ and $\div(r_2, s) = r_2(1/s)$, and so $\div(r_1, s)$ and $\div(r_2, s)$ can be viewed as different ``scalings" of the multiplicative inverse $1/s$ of s.

This is actually a special case of the previous example. In particular, we can define a function $f \in C(\mathbb{R}\setminus\{0\})$ by $f(s) = 1/s$, for each $s \in \mathbb{R}\setminus\{0\}$. Then, given any two real numbers $r_1, r_2 \in \mathbb{R}$, the functions $r_1 f$ and $r_2 f$ can be defined by,

$$r_1 f(\cdot) = \div(r_1,\cdot) \text{ and } r_2 f(\cdot) = \div(r_2,\cdot), \text{ respectively.}$$

Strictly speaking, there is nothing in the definition that precludes S from equaling $\mathbb{F}$. Consequently, addition, subtraction, and multiplication can all be seen as examples of scaling operations on $\mathbb{R}$. As with binary operations, it is easy to define any number of scaling operations upon a given nonempty set S. However, we are generally only interested in operations that are essentially like scalar multiplication on $\mathbb{R}^n$, and it is also quite common to additionally impose conditions for how scaling operations should interact with any binary operations that might also be defined upon S.

Put another way, the definitions for binary operation and scaling operation are not particularly useful when taken as is. Since these operations are allowed to be any functions having the proper domains, there is no immediate sense of meaningful abstraction. Instead, binary and scaling operations become useful when additionally conditions are placed upon them so that they can be used to abstract "arithmetic-like" properties. In other words, we are usually only interested in operations that abstract the salient properties of familiar operations for combining things like numbers, n-tuples, and functions.

Groups, Fields and Vector Spaces

Let G be a nonempty set, and let $*$ be a binary operation on G. (In other words, $*: G \times G \to G$ is a function with $*(a,b)$ denoted by $a * b$, for each $a, b \in G$.) Then G is said to form a group under $*$ if the following three conditions are satisfied:

- (Associativity) Given any three elements $a, b, c \in G$,

  $(a*b)*c = a*(b*c)$.

- (Existence of an identity element) There is an element $e \in G$ such that, given any element $a \in G$,

  $a*e = e*a = a$.

- (Existence of inverse elements) Given any element $a \in G$, there is an element $b \in G$ such that,

  $a*b = b*a = e$.

You should recognize these three conditions (which are sometimes collectively referred to as the group axioms) as properties that are satisfied by the operation of addition on $\mathbb{R}$. This is not an accident. In particular, given real numbers $\alpha, \beta \in \mathbb{R}$, the group axioms form the minimal set of assumptions needed in order to solve the equation $x + \alpha = \beta$ for the variable x, and it is in this sense that the group axioms are an abstraction of the most fundamental properties of addition of real numbers.

A similar remark holds regarding multiplication on $\mathbb{R}\setminus\{0\}$ and solving the equation $\alpha x=\beta$ for the variable x. Note, however, that this cannot be extended to all of $\mathbb{R}$. Because the group axioms are so general, they are particularly useful in building more complicated algebraic structures. This is done by adding any number of additional axioms, the most fundamental of which is as follows:

Let G be a group under binary operation $*$. Then G is called an abelian group (a.k.a. commutative group) if, given any two elements $a, b \in G$, $a*b = b*a$. Examples of groups are everywhere in abstract mathematics. We now give some of the more important examples that occur in Linear

Algebra. Please note, though, that these examples are primarily aimed at motivating the definitions of more complicated algebraic structures.

If $G \in \{\mathbb{Z}, \mathbb{Q}, \mathbb{R}, \mathbb{C}\}$, then G forms an abelian group under the usual definition of addition. The set $\mathbb{Z}_+$ of positive integers does not form a group under addition since, e.g., it does not contain an additive identity element.

Similarly, if $G \in \{\mathbb{Q}\backslash\{0\}, \mathbb{R}\backslash\{0\}, \mathbb{C}\backslash\{0\}\}$, then G forms an abelian group under the usual definition of multiplication. Note, though, that $\mathbb{Z}\backslash\{0\}$ does not form a group under multiplication since only ± 1 have multiplicative inverses.

If $m, n \in \mathbb{Z}_+$ are positive integers and $\mathbb{F}$ denotes either $\mathbb{R}$ or $\mathbb{C}$, then the set $F^{m\times n}$ of all $m \times n$ matrices forms an abelian group under matrix addition. Note, though, that $F^{m\times n}$ does not form a group under matrix multiplication unless $m = n = 1$, in which case $F^{1\times 1} = \mathbb{F}$.

Similarly, if $n \in \mathbb{Z}_+$ is a positive integer and $\mathbb{F}$ denotes either $\mathbb{R}$ or $\mathbb{C}$, then the set $GL(n, \mathbb{F})$ of invertible $n \times n$ matrices forms a group under matrix multiplications. This group, which is often called the general linear group, is non-abelian when $n \geq 2$. Note, though, that $GL(n, \mathbb{F})$ does not form agroup under matrix addition for any choice of n since, e.g., the zero matrix $0_{n\times n} \notin GL(n, \mathbb{F})$.

In the above examples, you should notice two things. First of all, it is important to specify the operation under which a set might or might not be a group. Second, and perhaps more importantly, all but one example is an abelian group. Most of the important sets in Linear Algebra possess some type of algebraic structure, and abelian groups are the principal building block of virtually every one of these algebraic structures. In particular, fields and vector spaces and rings and algebra can all be described as "abelian groups plus additional structure".

Given an abelian group G, adding "additional structure" amounts to imposing one or more additional operation on G such that each new operations is "compatible" with the preexisting binary operation on G. As our first example of this, we add another binary operation to G in order to obtain the definition of a field:

Let F be a nonempty set, and let $+$ and $*$ be binary operations on F. Then F forms a field under $+$ and $*$ if the following three conditions are satisfied:

- F forms an abelian group under $+$.
- Denoting the identity element for $+$ by 0, $F\backslash\{0\}$ forms an abelian group under $*$.
- ($*$ distributes over $+$) Given any three elements $a, b, c \in F$,
  $a*(b+c)=a*b+a*c.$

You should recognize these three conditions (which are sometimes collectively referred to as the field axioms) as properties that are satisfied when the operations of addition and multiplication are taken together on $\mathbb{R}$. This is not an accident. As with the group axioms, the field axioms form the minimal set of assumptions needed in order to abstract fundamental properties of these familiar arithmetic operations. Specifically, the field axioms guarantee that, given any field F, three conditions are always satisfied:

- Given any $a, b \in F$, the equation $x + a = b$ can be solved for the variable x.

- Given any $a \in F\backslash\{0\}$ and $b \in F$, the equation $a * x = b$ can be solved for x.
- The binary operation $*$ (which is like multiplication on $\mathbb{R}$) can be distributed over (i.e., is "compatible" with) the binary operation + (which is like addition on $\mathbb{R}$).

It should be clear that, if $F \in \{\mathbb{Q}, \mathbb{R}, \mathbb{C}\}$, then F forms a field under the usual definitions of addition and multiplication. Note, though, that the set $\mathbb{Z}$ of integers does not form a field under these operations since $\mathbb{Z}\backslash\{0\}$ fails to form a group under multiplication.

In some sense $\mathbb{Q}$, $\mathbb{R}$, and $\mathbb{C}$ are the only easily describable fields. While there are many other interesting and useful examples of fields, none of them can be described using entirely familiar sets and operations. This is because the field axioms are extremely specific in describing algebraic structure. We can build a much more general algebraic structure called a "ring" by still requiring that F form an abelian group under + but simultaneously relaxing the requirement that F simultaneously form an abelian group under $*$.

Rather than place an additional (and multiplication-like) binary operation on an abelian group, we instead impose a special type of scaling operation called scalar multiplication. In essence, scalar multiplication imparts useful algebraic structure on an arbitrary nonempty set S by indirectly imposing the algebraic structure of $\mathbb{F}$ as an abelian group under multiplication. (Recall that $\mathbb{F}$ can be replaced with either $\mathbb{R}$ or $\mathbb{C}$).

Let S be a nonempty set, and let $*$ be a scaling operation on S. (In other words, $*: \mathbb{F} \times S \to S$ is a function with $*(\alpha, s)$ denoted by $\alpha * s$ or even just αs, for every $\alpha \in \mathbb{F}$ and $s \in S$.) Then $*$ is called scalar multiplication if it satisfies the following two conditions:

- (Existence of a multiplicative identity element for $*$) Denote by 1 the multiplicative identity element for $\mathbb{F}$. Then, given any $s \in S$, $1 * s = s$.
- (Multiplication in $\mathbb{F}$ is quasi-associative with respect to $*$) Given any $\alpha, \beta \in \mathbb{F}$ and any $s \in S$,

  $(\alpha\beta) * s = \alpha * (\beta * s)$.

We choose to have the multiplicative part of $\mathbb{F}$ "act" upon S because we are abstracting scalar multiplication on both $\mathbb{R}^n$ and $C(\mathbb{R})$. This is because, by also requiring a "compatible" additive structure (called vector addition), we obtain the following alternate formulation for the definition of a vector space.

Let V be an abelian group under the binary operation +, and let $*$ be a scalar multiplication operation on V with respect to $\mathbb{F}$. Then V forms a vector space over $\mathbb{F}$ with respect to + and $*$ if the following two conditions are satisfied:

- ($*$ distributes over +) Given any $\alpha \in \mathbb{F}$ and any $u, v \in V$,

  $\alpha * (u + v) = \alpha * u + \alpha * v$.

- ($*$ distributes over addition in $\mathbb{F}$) Given any $\alpha, \beta \in \mathbb{F}$ and any $v \in V$,

  $(\alpha + \beta) * v = \alpha * v + \beta * v$.

Rings and Algebras

Specifically, we first "relax" the definition of a field in order to define a ring, and we then combine the definitions of ring and vector space in order to define an algebra. In some sense, groups, rings, and fields are the most fundamental algebraic structures, with vector spaces and algebras being particularly important variants within the study of Linear Algebra and its applications.

Let $\mathbb{R}$ be a nonempty set, and let $+$ and $*$ be binary operations on $\mathbb{R}$. Then $\mathbb{R}$ forms an (associative) ring under $+$ and $*$ if the following three conditions are satisfied:

- R forms an abelian group under $+$.
- ($*$ is associative) Given any three elements $a,b,c \in R$, $a*(b*c)=(a*b)*c$.
- ($*$ distributes over $+$) Given any three elements a, b, c $\in \mathbb{R}$,

  $a*(b+c)=a*b+a*c$ and $(a+b)*c=a*c+b*c$.

As with the definition of group, there are many additional properties that can be added to a ring; here, each additional property makes a ring more field-like in some way.

Let $\mathbb{R}$ be a ring under the binary operations $+$ and $*$. Then we call $\mathbb{R}$:

- Commutative: If $*$ is a commutative operation; i.e., given any $a,b \in R$, $a*b=b*a$.
- Unital: If there is an identity element for $*$; i.e., if there exists an element $i \in \mathbb{R}$ such that, given any $a \in R$, $a*i=i*a=a$.
- A commutative ring with identity if it's both commutative and unital.

In particular, note that a commutative ring with identity is almost a field; the only thing missing is the assumption that every element has a multiplicative inverse. It is this one difference that results in many familiar sets being CRIs (or at least unital rings) but not fields. E.g., $\mathbb{Z}$ is a CRI under the usual operations of addition and multiplication, yet, because of the lack of multiplicative inverses for all elements except ±1, $\mathbb{Z}$ is not a field.

In some sense, $\mathbb{Z}$ is the prototypical example of a ring, but there are many other familiar examples. E.g., if F is any field, then the set of polynomials F[z] with coefficients from F is a CRI under the usual operations of polynomial addition and multiplication, but again, because of the lack of multiplicative inverses for every element, F[z] is itself not a field. Another important example of a ring comes from Linear Algebra. Given any vector space V, the set $\mathcal{L}(V)$ of all linear maps from V into V is a unital ring under the operations of function addition and composition. However, $\mathcal{L}(V)$ is not a CRI unless $\dim(V) \in \{0,1\}$.

Alternatively, if a ring R forms a group under $*$ (but not necessarily an abelian group), then R is sometimes called a skew field (a.k.a. division ring). Note that a skew field is also almost a field; the only thing missing is the assumption that multiplication is commutative. Unlike CRIs, though, there are no simple examples of skew fields that are not also fields. As you can probably imagine, many other properties that can be appended to the definition of a ring, some of which are more useful than others.

Let A be a nonempty set, let + and × be binary operations on A, and let $*$ be scalar multiplication on A with respect to $\mathbb{F}$. Then A forms an (associative) algebra over $\mathbb{F}$ with respect to +, ×, and $*$ if the following three conditions are satisfied:

- A forms an (associative) ring under + and ×.
- A forms a vector space over $\mathbb{F}$ with respect to + and $*$.
- ($*$ is quasi-associative and homogeneous with respect to ×) Given any element $\alpha \in \mathbb{F}$ and any two elements a, b ∈ R,

  $\alpha*(a\times b)=(\alpha*a)\times b$ and $\alpha*(a\times b)=a\times(\alpha*b)$.

Two particularly important examples of algebras were already defined above: F[z] (which is unital and commutative) and $\mathcal{L}(V)$ (which is, in general, just unital). On the other hand, there are also many important sets in Linear Algebra that are not algebras. E.g., $\mathbb{Z}$ is a ring that cannot easily be made into an algebra, and $\mathbb{R}^3$ is a vector space but cannot easily be made into a ring (since the cross product operation from Vector Calculus is not associative).

Algebraic Functions

A function $y=f(x_1\ldots x_n)$ of the variables $x_1 \ldots x_n$ that satisfies an equation,

$$F(y,x_1\ldots x_n)=0,$$

where F is an irreducible polynomial in y, $x_1\ldots x_n$ with coefficients in some field K, known as the field of constants. The algebraic function is said to be defined over this field, and is called an algebraic function over the field K. The polynomial $F(y,x_1\ldots x_n)$ is often written in powers of the variable y, so that equation above assumes the form,

$$P_k(x_1\ldots x_n)y^k+P_{k-1}(x_1\ldots x_n)y^{k-1}+\cdots+$$
$$+P_0(x_1\ldots x_n)=0,$$

where $P_k(x_1\ldots x_n)\ldots P_0(x_1\ldots x_n)$ are polynomials in $x_1\ldots x_n$, and with $P_k(x_1\ldots x_n)\not\equiv 0$. The number k is the degree of F with respect to y, and is called the degree of the algebraic function. If k = 1, an algebraic function may be represented as a quotient,

$$y=-\frac{P_0(x_1\ldots x_n)}{P_1(x_1\ldots x_n)}$$

of polynomials, and is called a rational function of $x_1\ldots x_n$. For k=2, 3, 4, an algebraic function can be expressed as square and cube roots of rational functions in the variables $x_1\ldots x_n$; if k > 4, this is impossible in general.

The theory of algebraic functions was studied in the past from three different points of view: the function-theoretical point of view taken, in particular, by N.H. Abel, K. Weierstrass and B.

Riemann; the arithmetic-algebraic point of view taken by R. Dedekind, H. Weber and K. Hensel; and the algebraic-geometrical point of view, which originated with the studies of A. Clebsch, M. Noether and others. The first direction of the theory of algebraic functions of a single variable is connected with the study of algebraic functions over the field of complex numbers, in which they are regarded as meromorphic functions on Riemann surfaces and complex manifolds; the most important methods applied are the geometrical and topological methods of the theory of analytic functions. The arithmetic-algebraic approach involves the study of algebraic functions over arbitrary fields. The methods employed are purely algebraic.

The theory of valuations and extensions of fields are especially important. In the algebraic-geometrical approach algebraic functions are considered to be rational functions on an algebraic variety, and are studied by methods of algebraic geometry. These three points of view originally differed not only in their methods and their ways of reasoning, but also in their terminology. This differentiation has by now become largely arbitrary, since function-theoretical studies involve the extensive use of algebraic methods, while many results obtained at first using function-theoretical and topological methods can be successfully applied to more general fields using algebraic analogues of these methods.

Algebraic Functions of One Variable

Over the field C of complex numbers, an algebraic function of one variable y = f(x)(or y(x) for short) is a k-valued analytic function. If D(x) is the discriminant of the polynomial,

$$F(x,y)=P_k(x)y^k+\cdots+P_1(x)y+P_0(x),$$
$$P_k(x)\neq 0,$$

(i.e. of the polynomial for which F(x, f(x))=0), which is obtained by eliminating y from the equations,

$$F(x,y)=0, \frac{\partial F(x,y)}{\partial y}=0,$$

to yield the equation,

$$P_k(x)D(x)=0,$$

then the roots $x_1 \ldots x_m$ of this last equation are known as the critical values of y = f(x). The complementary set G = C\{$x_1 \ldots x_m$} is known as the non-critical set. For any point $x_0 \in G$ equation $F(x,y)=P_k(x)y^k+\cdots+P_1(x)y+P_0(x), P_k(x)\neq 0,$ has k different roots $y_0^1 \ldots y_0^k$ and the condition,

$$\frac{\partial F(x_0,y_0^j)}{\partial y}\neq 0,\ j=1\ldots k,$$

is satisfied. According to the implicit function theorem, in a neighbourhood of the point x_0 there exist k single-valued analytic functions $f_0^1(x)\ldots f_0^k(x)$ which satisfy the conditions,

$$f_0^j(x_0)=y_0^j,\ F(x,f_0^j(x))=0,$$

and which can be decomposed into a convergent series,

$$f_0^j(x)=y_0^j+\alpha_1^j(x-x_0)+\alpha_2^j(x-x_0)^2+\dots$$

Thus, for each point $x_0 \in G$ one can construct k elements of an analytic function, known as the function elements with centre at the point x_0. For any two points $x_1, x_2 \in G$, any elements $f_1^i(x)$ and $f_2^i(x)$ with centres at x_1 and x_2, respectively, are derived from each other by analytic continuation along some curve in G; in particular, any two elements with the same centres are also connected in this way. If x_0 is a critical point of an algebraic function, then two cases are possible: 1) x_0 is a root of the discriminant, i.e. $D(x_0)=0$ but $P_k(x_0)\neq 0$; or 2) $P_k(x_0)=0$.

Case: Let K_0 be a small disc with centre at x_0 which does not contain other critical points, and let $f_1'(x)\dots f_k'(x)$ be a system of regular elements with centre at $x' \in K_0$, $x' \neq x_0$. These functions remain bounded as $x \to x_0$. Furthermore, let D be the circle with centre x_0 passing through x'; it is completely contained inside K_0. The analytic continuation of some given element, e.g. $f_1'(x)$ along D(in, say, the clockwise direction), yields an element $f'(x)$ which also belongs to the system of elements with centre x'. This system consists of k elements, and a minimum required finite number $a_1 \leq k$ of such turns yields the initial element $f_1'(x)$. One obtains a subsystem $f_1'(x)\dots f_{a_1}'(x)$ of elements with centre x'; each one of these elements may be obtained by analytic continuation of the other by a number of turns around the point x_0; such a subsystem is known as a cycle. Any system $f_1(x)\dots f_k(x)$ can be decomposed into a number of non-intersecting cycles,

$$\{f_1'(x)\dots f_{a_1}'(x)\},$$

$$\{f_{a_1}'+1(x)\dots f_{a_1+a_2}'(x)\}\dots$$

$$\{f_{a_1+\dots+a_{s-1}}'+1(x)\dots f_{a_1+\dots+a_s}'(x)\},$$

$a_1+\dots+a_s=k$. If $a_1 > 1$, then the element $f_1'(x)$ is not a single-valued function of x in the disc K_0, but is a single-valued analytic function of the parameter $\tau=(x-x_0)^{1/a_1}$ in a neighbourhood of $\tau = 0$. In a certain neighbourhood of this point the elements $f_1'(x)\dots f_{a_1}'(x)$ of the first cycle can be represented as convergent series,

$$f_1'(x)=\sum_{i=0}^{\infty}\alpha_i^1\tau^i=\sum_{i=0}^{\infty}\alpha_i^1(x-x_0)^{i/a_1},$$

....

$$f_{\alpha_1}'(x)=\sum_{i=0}^{\infty}\alpha_i^{a_1}\tau^i=\sum_{i=0}^{\infty}\alpha_i^{a_1}(x-x_0)^{i/a_1},$$

and similar expansions also take place for the elements of other cycles. Such expansions of elements by fractional degrees of the difference $x - x_0$, where x_0 is a critical point, are known as Puiseux series. The transformation $\tau \to \tau_r$, $r = e^{2\pi i/a}$, which corresponds to one turn around x_0, converts the Puiseux series of elements in one cycle into each other in cyclic order, i.e. there is a cyclic permutation of the series and of the corresponding elements. To turns around the critical point correspond permutations of the elements with centre at this point; these permutations consist of cycles of the orders $a_1\dots a_s$, $\sum a_i=k$. The permutations defined in this way constitute the monodromy group of the algebraic function. If at least one element a_i is greater than 1, the critical point x_0 is called an algebraic branch point of the algebraic function; the numbers a_i(sometimes a_i-1) are called the branch indices (or branch orders) of the algebraic function.

Case: If $P_k(x)y$ is substituted for y, one returns to the previous case; one obtains expansions similar to the above equation, which may contain a finite number of terms with negative indices:

$$f(x)=\sum_{i=-p}^{\infty}\alpha_i\tau^i=\sum_{i=-p}^{\infty}\alpha_i(x-x_0)^{i/a}.$$

If $p > 0$, the point x_0 is a pole of order p of the algebraic function. An algebraic function is usually considered on the Riemann sphere S, i.e. on the complex plane completed by the point at infinity $x = \infty$. The introduction of the variable $\tau = 1/x$ reduces this case to the previous case; in a neighbourhood of the point $\tau = 0(x = \infty)$ one has the expansion,

$$y(x)=\sum_{j=-r}^{\infty}\alpha_j\tau^{j/a}=\sum_{j=-r}^{\infty}\alpha_j x^{-j/a}.$$

If $r > 0$, then the point $x=\infty$ is called a pole of order r.

The parameter of the expansion in the series of above four equations is called the local uniformizing parameter for the algebraic function. If x_0 is a non-critical point of the algebraic function, then $\tau = x - x_0$ can be taken as parameter; if, on the other hand, x_0 is a critical point, the root $(x-x_0)^{1/a}$(where a is a positive integer) can be taken as such a parameter. The population of all elements of an algebraic function described above forms the complete algebraic function in the sense of Weierstrass. Algebraic functions have no singularities other than algebraic branch points and poles. The converse proposition is also true: A function $y = f(x)$ which is analytic, is not more than s- valued at all points of the Riemann sphere except for a finite number of points $x_1...x_m$ and $x = \infty$, and has at such points only poles or algebraic branch points, is an algebraic function of degree $k \leq s$.

The Riemann surface of a complete algebraic function is compact and is a k-sheeted covering of the Riemann sphere, branch points can be the critical points and the point $x = \infty$. Algebraic functions are the only class of functions with a compact Riemann surface. The genus of the Riemann surface of an algebraic function is important; it is called the genus of the algebraic function. It can be calculated by the Riemann–Hurwitz formula. The genus of a rational function is zero, and its Riemann surface is the Riemann sphere. The Riemann surface of an elliptic function that satisfies a third- or fourth-degree equation is a torus; the genus of such a function is one.

The universal covering Riemann surface of an algebraic function is a simply-connected two-dimensional manifold, i.e. it has a trivial fundamental group and is conformally equivalent either to the Riemann sphere, the complex plane or the interior of the unit disc. In the first case the algebraic function is a rational, in the second case it is an elliptic, while in the third case it is a general function.

The uniformization problem of algebraic functions is closely connected with the theory of Riemann surfaces of algebraic functions. The function y = f(x) can be uniformized if y and x are representable as single-valued analytic functions,

$$y = y(t),\ x = x(t),$$

of a parameter t, which identically satisfy equation $F(x,y) = P_k(x)y^k + \cdots + P_1(x)y + P_0(x),\ P_k(x) \neq 0,$ The uniformization problem is locally solved by a local uniformizing parameter; however, it is the solution "in the large" that is of interest. If k = 1, i.e. if y(x) is a rational function of x, this parameter may be the variable $x - x_0$; if k = 2, then uniformization is attained with the aid of a rational or a trigonometric function. For instance, if y(x) satisfies the equation,

$$y^2 - x^2 = 1,$$

one can take,

$$y = \frac{t^2+1}{t^2-1},\ x = \frac{2t}{t^2-1},$$

or

$$y = \sec t, x = \operatorname{tg} t.$$

If k = 3, 4 in the case of an algebraic function of genus one, uniformization is achieved using elliptic functions. Finally, if k > 4 and the genus of the algebraic function is higher than one, uniformization is realized using automorphic functions.

Algebraic Functions of Several Variables

If f is an algebraic function in the variables $x_1 \dots x_n$, then the set of all rational functions $R(y, x_1 \dots x_n)$ forms a field K_f, coinciding with the field of rational functions on the algebraic hypersurface in (n+1)- dimensional space defined by the equation $F(y, x_1 \dots x_n) = 0$. If the field of constants k is the field of complex numbers C and if n = 1, then K_f is identical with the field of meromorphic functions on the Riemann surface of the algebraic function. The field K_f is an extension of finite type of the field of constants k of transcendence degree n. In particular, any n+1 elements of this field are connected by an algebraic equation, so that each of them defines an algebraic function of the remaining elements. Any extension K of finite type of a field k of transcendence degree n is known as an algebraic function field in n variables (or, sometimes, as a function field).

Each such field contains a purely transcendence extension $k(x_1 \dots x_n)$ of the field k(called the field of rational functions in n variables). Any element $y \in K$ satisfies some algebraic equation $\Phi(y, x_1 \dots x_n)=0$, and can be considered as an algebraic function in the variables $x_1 \dots x_n$. Each field K of algebraic functions in n variables is isomorphic to the field of rational functions on some algebraic variety of dimension n, which is called a model of K. If the field of constants k is algebraically closed and of characteristic zero, then each algebraic function field has a non-singular projective model (cf. Resolution of singularities). Let S be the set of all non-trivial valuations of an algebraic function field K which are non-negative on the field of constants. If provided with the natural topology, it is known as the abstract Riemann surface of the field K. In the case of algebraic functions in one variable, the Riemann surface coincides with the set of non-singular projective models, which in this case is uniquely defined up to an isomorphism. Many concepts and results in algebraic geometry on the model of a field K can be restated in the language of the theory of valuations of fields. A particularly close analogy holds for algebraic functions in one variable, the theory of which is practically identical with the theory of algebraic curves.

Each algebraic function field in one variable is the field of fractions of a Dedekind ring, so that many results and concepts of the theory of divisibility in algebraic number fields can be applied to function fields. Many problems and constructions in algebraic number theory motivate similar problems and constructions in fields of algebraic functions and vice versa. For instance, the application of a Puiseux expansion to the theory of algebraic numbers led to the genesis of the p-adic method in number theory, due to Hensel. Class field theory, which had originally belonged to the domain of algebraic numbers, was subsequently applied to functions. An especially close analogy exists between algebraic number fields and algebraic function fields over a finite field of constants. For instance, the concept of a zeta-function is defined for the latter and the analogue of the Riemann hypothesis has been demonstrated for algebraic function fields.

Algebraic Expressions

An algebraic expression is a mathematical term that consists of variables and constants along with mathematical operators (subtraction, addition, multiplication, etc). Example: $8x - 20$, $5x - 6y + 30$ etc.

Algebraic expressions shouldn't be confused with algebraic equations. An algebraic equation has two sides (Left Hand Side or LHS and Right Hand Side or RHS) whereas an algebraic expression does not. In fact, an algebraic equation is made of two or more algebraic expressions separated by an equal sign (=). Let's understand the difference with an example. For example: $(a + b)^2 = a^2 + 2ab + b^2$ is an algebraic equation containing two terms, whereas $(a + b)^2$ and $(a^2 + 2ab + b^2)$ are algebraic expressions.

Parts of an Algebraic Expression

An algebraic expression contains a variable with or without a coefficient, a mathematical operator, and sometimes a constant. The different parts of an algebraic expression are:

- Variable: A variable is a letter whose value is unknown. It can take any value depending upon the situation.
- Coefficient: The coefficient is a numerical value used before a variable to modify its value. It may or may not be present in an algebraic expression.
- Constant: A constant is any term whose value remains unchanged throughout the algebraic expression.
- Operator: Mathematical operators are used in an algebraic expression to perform some mathematical calculation on two or more expressions.

Let's understand with an example:

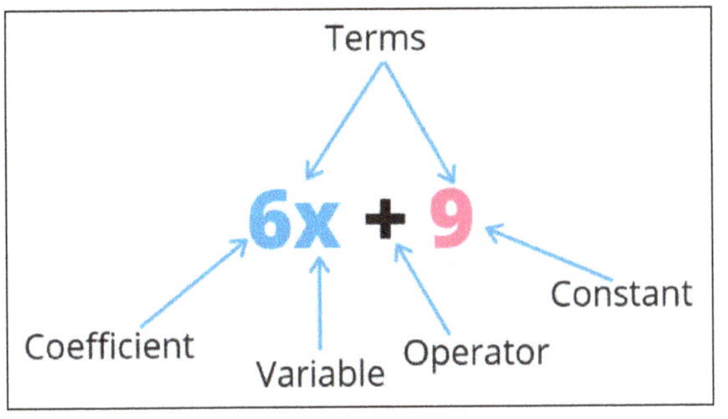

Here:

- 6 is the coefficient of x.
- x is a variable. Its value unknown and can be anything.
- 9 is a constant with a fixed value.
- + is an operator that is used for addition.

The expression as a whole is a binomial term since it contains only two terms i.e. 6x and 9.

Algebraic Expressions and Terms

Here we will tell you the difference between a variable, expression, and a term and how they are related. Often in mathematics, we come across symbols or letters such as x, y, z, etc. whose value is unknown and we have to find the exact or approximate value. Hence, a variable is defined as a letter or a symbol that is used to represent an unknown number Any number or a variable combined by multiplication or division is known as a term. Example: 5, 6x, 5/3, etc.

An expression is a collection of a single or multiple terms separated by either subtraction or addition. Example: 5x-3, 23x, 2/3x + 4.

Algebraic Expressions and its Types

There are six main types of algebraic expressions. These are as under:

- Monomial Expression: An expression with a single term is known as a monomial. $8x^6$, 10xy, 12xyz, etc. are examples of monomial expressions.
- Binomial Expression: An expression with two terms is known as a binomial. $8x^6 + 3$, $10xy - x^3$, $12xy + 4$, etc. are examples of Binomial expressions.
- Trinomial Expression: An expression with three terms is known as a trinomial. $x + x^2 + \pi$, $y^4 + y + 6$, $z + z^2 - 5$, etc. are examples of trinomial expressions.
- Linear Polynomial: A polynomial of degree 1 is called a linear polynomial. In other words, in a linear polynomial, the highest exponent of the variable is one. Example: (2x + 1), (8 – 2u), etc.

- Quadratic Polynomial: A polynomial of degree 2 is called a quadratic polynomial. This means, in a quadratic polynomial, the highest exponent of the variable is two. Example: $x^2 + 25$, $y^2 - y - 36$, etc.
- Cubic Polynomial: A polynomial of degree 3 is called a linear polynomial. Here, the highest exponent of the variable is three. Example: $x^3 + x^2$, $x^3 + 5$, etc.

Some other types of algebraic expressions are:

- Numeric Expression: A numerical expression consists of only numbers and mathematical operators. There is no variable present in a numeric expression. For example: 5+14, 9+6, 23, etc.
- Variable Expression: A variable expression consists only of variables and mathematical operators. The variables may or may not contain coefficients. For example: 5x+3y, 8x+z, a+b, etc.

Now that you are familiar with all the types of algebraic expressions, let's have a look at how to simplify them.

Simplification of Algebraic Expressions

For this, we will take a polynomial, $P(x) = 5x^2 - 3x + 7$ and we have to simplify this polynomial for x = 2. Here, we'll solve this step by step.

- 1st Step: Put the value of x = 2.
- $P(2) = 5(2)^2 - 3(2) + 7$
- 2nd Step: Use BODMAS method,
- $P(2) = 5(2 * 2) - 3 * 2 + 7$
- 3rd Step: The expression will become,

  $P(2) = 5(4) - 6 + 7$

  $P(2) = 5 * 4 - 6 + 7$

  $P(2) = 20 - 6 + 7$

  $P(2) = 27 - 6$

  $P(2) = 21.$

That was the process for simplifying an algebraic expression step by step.

Quadratic Equation

A quadratic equation is a second-order polynomial equation in a single variable x,

$$ax^2 + bx + c = 0,$$

with $a \neq 0$. Because it is a second-order polynomial equation, the fundamental theorem of algebra guarantees that it has two solutions. These solutions may be both real, and both complex.

The roots x can be found by completing the square,

$$x^2 + \frac{b}{a}x = -\frac{c}{a}$$

$$\left(x + \frac{b}{2a}\right)^2 = -\frac{c}{a} + \frac{b^2}{4a^2} = \frac{b^2 - 4ac}{4a^2}$$

$$x + \frac{b}{2a} = \frac{\pm\sqrt{b^2 - 4ac}}{2a}.$$

Solving for x then gives,

$$x = \frac{-b \pm \sqrt{b^2 - 4ac}}{2a}.$$

This equation is known as the quadratic formula.

The first known solution of a quadratic equation is the one given in the Berlin papyrus from the Middle Kingdom in Egypt. This problem reduces to solving,

$$x^2 + y^2 = 100$$

$$y = \frac{3}{4}x.$$

The Greeks were able to solve the quadratic equation by geometric methods, and Euclid's Data contains three problems involving quadratics. In his work Arithmetica, the Greek mathematician Diophantus solved the quadratic equation, but giving only one root, even when both roots were positive.

A number of mathematicians gave rules equivalent to the quadratic formula. It is possible that certain altar constructions dating from ca. 500 BC represent solutions of the equation, but even should this be the case, there is no record of the method of solution. The Hindu mathematician Āryabhata gave a rule for the sum of a geometric series that shows knowledge of the quadratic equations with both solutions.

An alternate form of the quadratic equation is given by dividing (◇) through by x^2:

$$a + \frac{b}{x} + \frac{c}{x^2} = 0$$

$$c\left(\frac{1}{x^2} + \frac{b}{cx}\right) + a = 0$$

$$c\left(\frac{1}{x}+\frac{b}{2c}\right)^2 = c\left(\frac{b}{2c}\right)^2 - a = \frac{b^2}{4c} - \frac{4ac}{4c} = \frac{b^2-4ac}{4c}.$$

Therefore,

$$\frac{1}{x}+\frac{b}{2c} = \pm\frac{\sqrt{b^2-4ac}}{2c}$$

$$\frac{1}{x} = \frac{-b\pm\sqrt{b^2-4ac}}{2c}$$

$$x = \frac{2c}{-b\pm\sqrt{b^2-4ac}}.$$

This form is helpful if $b^2 >> 4ac$, where >> denotes much greater, in which case the usual form of the quadratic formula can give inaccurate numerical results for one of the roots. This can be avoided by defining,

$$q \equiv -\frac{1}{2}\left[b+\operatorname{sgn}(b)\sqrt{b^2-4ac}\right].$$

so that b and the term under the square root sign always have the same sign. Now, if b > 0, then,

$$q = -\frac{1}{2}\left(b+\sqrt{b^2-4ac}\right)$$

$$\frac{1}{q} = \frac{-2}{b+\sqrt{b^2-4ac}}\frac{b-\sqrt{b^2-4ac}}{b-\sqrt{b^2-4ac}} = \frac{-2\left(b-\sqrt{b^2-4ac}\right)}{b^2-\left(b^2-4ac\right)}$$

$$= \frac{-2\left(b-\sqrt{b^2-4ac}\right)}{4ac} = \frac{-b+\sqrt{b^2-4ac}}{2ac},$$

so,

$$x_1 \equiv \frac{q}{a} = \frac{-b-\sqrt{b^2-4ac}}{2a}$$

$$x_2 \equiv \frac{c}{q} = \frac{-b+\sqrt{b^2-4ac}}{2a}.$$

Similarly, if b < 0, then,

$$q = -\frac{1}{2}\left(b-\sqrt{b^2-4ac}\right) = \frac{1}{2}\left(-b+\sqrt{b^2-4ac}\right)$$

$$\frac{1}{q}=\frac{2}{-b+\sqrt{b^2-4ac}}\frac{b+\sqrt{b^2-4ac}}{b+\sqrt{b^2-4ac}}=\frac{2\left(b+\sqrt{b^2-4ac}\right)}{-b^2+\left(b^2-4ac\right)}$$

$$=\frac{b+\sqrt{b^2-4ac}}{-2ac}=\frac{-b-\sqrt{b^2-4ac}}{2ac},$$

so,

$$x_1\equiv\frac{q}{a}=\frac{-b+\sqrt{b^2-4ac}}{2a}$$

$$x_2\equiv\frac{c}{q}=\frac{-b-\sqrt{b^2-4ac}}{2a}.$$

Therefore, the roots are always given by $x_1=q/a$ and $x_2=c/q$.

Now consider the equation expressed in the form,

$$a_2x^2+a_1x+a_0=0,$$

with solutions z_1 and z_2. These solutions satisfy Vieta's formulas,

$$z_1+z_2=-\frac{a_1}{a_2}$$

$$z_1z_2=\frac{a_0}{a_2}.$$

The properties of the symmetric polynomials appearing in Vieta's formulas then give,

$$z_1^2+z_2^2=\frac{a_1^2-2a_0a_2}{a_2^2}$$

$$z_1^3+z_2^3=-\frac{a_1^3-3a_0a_1a_2}{a_2^3}$$

$$z_1^4+z_2^4=\frac{a_1^4-4a_0a_1^2a_2+2a_0^2a_2^2}{a_2^4}.$$

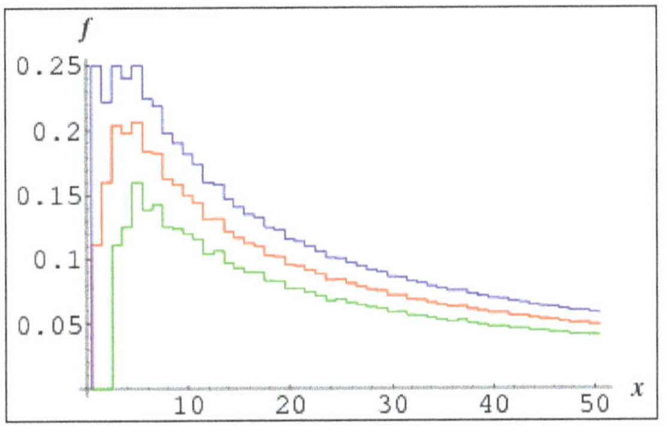

Given a quadratic integer polynomial $x^2 + ax + b$, consider the number of such polynomials that are factorable over the integers for a and b taken from some set of integers $S \subset \mathbb{Z}$. For example, for $S = \{1, 2, 3, 4\}$, there are four such polynomials,

$$x^2 + 2x + 1 = (x+1)^2$$

$$x^2 + 3x + 2 = (x+1)(x+2)$$

$$x^2 + 4x + 3 = (x+1)(x+3)$$

$$x^2 + 4x + 4 = (x+2)^2$$

The following table summarizes the counts of such factorable polynomials for simple $S \subset \mathbb{Z}$ and small n. Plots of the fractions of factorable polynomials for $S = \{-n, ..., n-1, n\}$ (red), $S = \{0, 1, ..., n\}$ (blue), and $S = \{1, 2, ..., n\}$ (green). Amazingly, the sequence for [-n, n] has the recurrence equation,

$$a_n = a_{n-1} + 2[d(n)+1] + \mathcal{X}_{n^2}(n)$$

where $d(n) = \sigma_0(n)$ is the number of divisors of n and $\mathcal{X}_{n^2}(n)$ is the characteristic function of the square numbers.

$S \subset \mathbb{Z}$	OEIS	factorable over S for $n = 0, 1, ...$
$[-n,n]$	A067274	1, 4, 10, 16, 25, 31, 41, 47, 57, ...
$[0,n]$	A091626	1, 2, 4, 6, 9, 11, 14, 16, 19, 22, ...
$[1,n]$	A091627	0, 0, 1, 2, 4, 5, 7, 8, 10, 12, 14, ..

References

- What-is-importance-of-algebra-in-real-life-algebra-uses: ziyyara.com, Retrieved 09, July 2020
- Branches, algebra, maths: byjus.com, Retrieved 25, March 2020
- Linear-Algebra, A-Linear-Algebra-(Schilling-Nachtergaele-and-Lankham), A-Summary-of-Algebraic-Structures: math.libretexts.org, Retrieved 30, August 2020
- Algebraic-expressions, exams: embibe.com, Retrieved 19, June 2020
- Quadratic-Equation: mathworld.wolfram.com, Retrieved 16, May 2020

2

Abstract Algebra

The study of algebraic structures is known as abstract algebra. Algebraic structures include rings, groups, modules, fields, vector spaces and lattices. Calculations that combine two elements are called binary operations. This chapter has been carefully written to provide an easy understanding of the varied facets of abstract algebra.

Abstract algebra is a broad field of mathematics, concerned with algebraic structures such as groups, rings, vector spaces, and algebras. Roughly speaking, abstract algebra is the study of what happens when certain properties of number systems are abstracted out; for instance, altering the definitions of the basic arithmetic operations result in a structure known as a ring, so long as the operations are consistent.

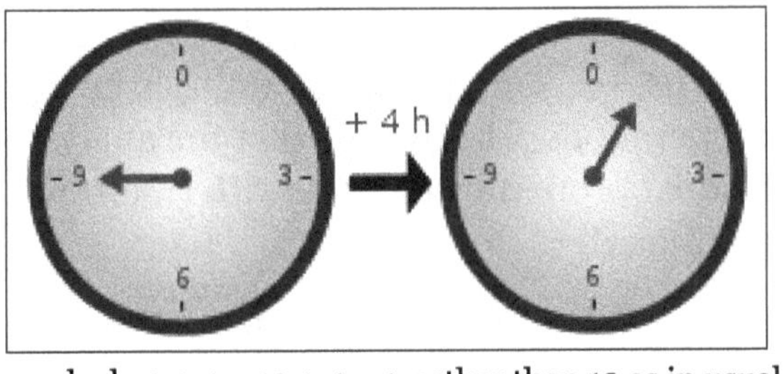

On the 12-hour clock, 9 + 4 = 19 + 4 = 1, rather than 13 as in usual arithmetic.

For example, the 12-hour clock is an example of such an object, where the arithmetic operations are redefined to use modular arithmetic (with modulus 12). An even further level of abstraction--where only one operation is considered--allows the clock to be understood as a group. In either case, the abstraction is useful because many properties can be understood without needing to consider the specific structure at hand, which is especially important when considering the relationship(s) between structures; the concept of a group isomorphism is an example.

Levels of Abstraction in Abstract Algebra

It is possible to abstract away practically all of the properties found in the "usual" number systems, the tradeoff being that the resulting object--known as a magma (which consists of a set and a binary operation, that need not satisfy any properties other than closure)--is simply too general to be interesting. On the other extreme, it is possible to abstract out practically no properties, which allows for many results to be found, but the resulting object (the usual number systems) is too specific to solve more general problems.

Most of abstract algebra is dedicated to objects that have a reasonable balance between generality and structure, most notably groups and rings in which most of the basic properties of arithmetic are maintained, but their specifics are left free. Still, some higher levels of abstraction are occasionally useful; quasigroups, for instance, are related to Latin squares, and monoids are often used in computer science and are simple examples of categories.

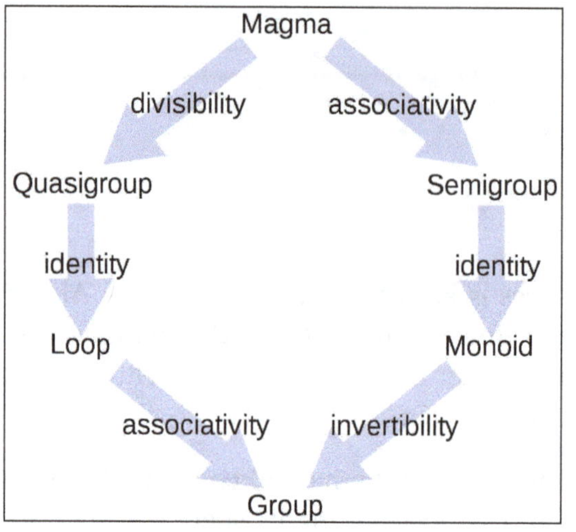

Ring Theory

Rings are one of the lowest level of abstraction, essentially obtained by overwriting the addition and multiplication functions simultaneously (compared to groups, which uses only one operation). Thus a ring is--in some sense--a combination of multiple groups, as a ring can be viewed as a group over either one of its operations. This means that the analysis of groups is also applicable to rings, but rings have additional properties to work with (the tradeoff being that rings are less general and require more conditions).

The definition of a ring is similar to that of a group, with the extra condition that the distributive law holds as well: A ring is a set R together with two operations + and · satisfying the following properties (ring axioms):

- R is an abelian group under addition. That is, R is closed under addition, there is an additive identity (called 0), every element $a \in R$ has an additive inverse $-a \in R$ and addition is associative and commutative.
- R is closed under multiplication, and multiplication is associative: $\forall a, b \in R$ a.b $\in$ R

  $\forall a,b,c \in R \;\; a\cdot(b\cdot c) = (a\cdot b)\cdot c.$
- Multiplication distributes over addition: $\forall a,b,c \in R$

  $a\cdot(b+c) = a\cdot b + a\cdot c$ and $(b+c)\cdot a = b\cdot a + c\cdot a.$

A ring is usually denoted by (R, +,.) and often it is written only as R when the operations are understood. For example, the integers $\mathbb{Z}$ form a ring, as do the integers modulo n ((denoted by Z_n). Less obviously, the square matrices of a given size also form a ring; this ring is noncommutative. Commutative ring theory, or commutative algebra, is much better understood than noncommutative rings are. As in groups, a ring homomorphism can be defined as a mapping preserving the structure of both operations.

Rings are used extensively in algebraic number theory, where "integers" are reimagined as slightly different objects (for example, Gaussian integers), and the effect on concepts such as prime factorization is analyzed. Of particular interest is the fundamental theorem of arithmetic, which involves the concept of unique factorization; in other rings, this may not hold, such as,

$$6 = 2 \cdot 3 = \left(1+\sqrt{-5}\right)\left(1-\sqrt{-5}\right).$$

Theory developed in this field solves problems ranging from sum of squares theorems to Fermat's last theorem.

Other Applications of Abstract Algebra

Abstract algebra also has heavy application in physics and computer science through the analysis of vector spaces. For example, the Fourier transform and differential geometry both have vector spaces as their underlying structures; in fact, the Poincare conjecture is (roughly speaking) a statement about whether the fundamental group of a manifold determines if the manifold is a sphere.

Related to vector spaces are modules, which are essentially identical to vector spaces but defined over a ring rather than over a field (and are thus more general). Modules are heavily related to representation theory, which views the elements of a group as linear transformations of a vector space; this is desirable to make an abstract object (a group) somewhat more concrete, in the sense that the group is better understood by translating it into a well-understood object in linear algebra (as matrices can be viewed as linear transformations, and vice versa). The relationships between various algebraic structures are formalized using category theory.

Sets

Sets, in mathematics, are an organized collection of objects and can be represented in set-builder form or roster form. Usually, sets are represented in curly braces {}, for example, A = {1, 2, 3, 4} is a set. In sets theory, you will learn about sets and its properties. It was developed to describe the collection of objects. You have already learned about the classification of sets here. The set theory defines the different types of sets, symbols and operations performed.

Sets are represented as a collection of well-defined objects or elements and it does not change from person to person. A set is represented by a capital letter. The number of elements in the finite set is known as the cardinal number of a set.

What are the Elements of a Set

Let us take an example: A = {1, 2, 3, 4, 5}

Since a set is usually represented by the capital letter. Thus, A is the set and 1, 2, 3, 4, 5 are the elements of the set or members of the set. The elements that are written in the set can be in any order but cannot be repeated. All the set elements are represented in small letter in case of alphabets. Also, we can write it as $1 \in A$, $2 \in A$ etc. The cardinal number of the set is 5. Some commonly used sets are as follows:

- N: Set of all natural numbers.
- Z: Set of all integers.

- Q: Set of all rational numbers.
- R: Set of all real numbers.
- Z^+: Set of all positive integers.

Order of Sets

The order of a set defines the number of elements a set is having. It describes the size of a set. The order of set is also known as the cardinality. The size of set whether it is a finite set or an infinite set said to be set of finite order or infinite order, respectively.

Representation of Sets

The sets are represented in curly braces, {}. For example, {2, 3, 4} or {a, b, c} or {Bat, Ball, Wickets}. The elements in the sets are depicted in either the Statement form, Roster Form or Set Builder Form.

Statement Form

In statement form, the well-defined descriptions of a member of a set are written and enclosed in the curly brackets. For example, the set of even numbers less than 15. In statement form, it can be written as {even numbers less than 15}.

Roster Form

In Roster form, all the elements of a set are listed. For example, the set of natural numbers less than 5.

Natural Number = 1, 2, 3, 4, 5, 6, 7, 8,......

Natural Number less than 5 = 1, 2, 3, 4

Therefore, the set is N = {1, 2, 3, 4}

Set Builder Form

The general form is, A = {x : property}.

Example: Write the following sets in set builder form: A = {2, 4, 6, 8}.

Solution:

$2 = 2 \times 1$

$4 = 2 \times 2$

$6 = 2 \times 3$

$8 = 2 \times 4$

So, the set builder form is $A = \{x: x=2n, n \in N \text{ and } 1 \le n \le 4\}$. Also, Venn Diagrams are the simple and best way for visualized representation of sets.

Types of Sets

We have several types of sets in Maths. They are empty set, finite and infinite sets, proper set, equal sets, etc. Let us go through the classification of sets here:

- Empty Set: A set which does not contain any element is called an empty set or void set or null set. It is denoted by { } or Ø. A set of apples in the basket of grapes is an example of an empty set because in a grapes basket there are no apples present.
- Singleton Set: A set which contains a single element is called a singleton set. Example: There is only one apple in a basket of grapes.
- Finite set: A set which consists of a definite number of elements is called a finite set. Example: A set of natural numbers up to 10.

  A = {1, 2, 3, 4, 5, 6, 7, 8, 9, 10}.
- Infinite set: A set which is not finite is called an infinite set. Example: A set of all natural numbers.

  A = {1,2,3,4,5,6,7,8,9……}
- Equivalent set: If the number of elements is the same for two different sets, then they are called equivalent sets. The order of sets does not matter here. It is represented as: n(A) = n(B) where A and B are two different sets with the same number of elements.

  Example: If A = {1, 2, 3, 4} and B = {Red, Blue, Green, Black}.

  In set A, there are four elements and in set B also there are four elements. Therefore, set A and set B are equivalent.
- Equal sets: The two sets A and B are said to be equal if they have exactly the same elements, the order of elements do not matter. Example: A = {1, 2, 3, 4} and B = {4, 3, 2, 1}.

  A = B
- Disjoint Sets: The two sets A and B are said to be disjoint if the set does not contain any common element. Example: Set A = {1, 2, 3, 4} and set B = {5, 6, 7, 8} are disjoint sets, because there is no common element between them.
- Subsets: A set 'A' is said to be a subset of B if every element of A is also an element of B, denoted as $A \subseteq B$. Even the null set is considered to be the subset of another set. In general, a subset is a part of another set.

  Example: A = {1, 2, 3}

  Then $\{1, 2\} \subseteq A$.

  Similarly, other subsets of set A are: {1},{2},{3},{1,2},{2,3},{1,3},{1,2,3},{}.

  The set is also a subset of itself. If A is not a subset of B, then it is denoted as $A \not\subset B$.
- Proper Subset: If $A \subseteq B$ and $A \neq B$, then A is called the proper subset of B and it can be

written as A⊂B. Example: If A = {2, 5, 7} is a subset of B = {2, 5, 7} then it is not a proper subset of B = {2, 5, 7}. But, A = {2, 5} is a subset of B = {2, 5, 7} and is a proper subset also.

- Superset: If set A is a subset of set B and all the elements of set B are the elements of set A, then A is a superset of set B. It is denoted by A⊃B. Example: If Set A = {1, 2, 3, 4} is a subset of B = {1, 2, 3, 4}. Then A is superset of B.

- Universal Set: A set which contains all the sets relevant to a certain condition is called the universal set. It is the set of all possible values. Example: If A = {1, 2, 3} and B {2, 3, 4, 5}, then universal set here will be: U = {1, 2, 3, 4, 5}.

Operations on Sets

In set theory, the operations of the sets are carried when two or more sets combine to form a single set under some of the given conditions. The basic operations on sets are:

- Union of sets.
- Intersection of sets.
- A complement of a set.
- Cartesian product of sets.
- Set difference.

Union of Sets

If set A and set B are two sets, then A union B is the set that contains all the elements of set A and set B. It is denoted as A ∪ B.

Example: Set A = {1, 2, 3} and B = {4, 5, 6}, then A union B is:

A ∪ B = {1, 2, 3, 4, 5, 6}.

Intersection of Sets

If set A and set B are two sets, then A intersection B is the set that contains only the common elements between set A and set B. It is denoted as A ∩ B.

Example: Set A = {1, 2, 3} and B = {4,5,6}, then A intersection B is:

A ∩ B = { } or Ø.

Since A and B do not have any elements in common, so their intersection will give null set.

Complement of Sets

The complement of any set, say P, is the set of all elements in the universal set that are not in set P. It is denoted by P’.

Properties of Complement sets:

- $P \cup P' = U$.
- $P \cap P' = \Phi$.
- Law of double complement : $(P')' = P$.
- Laws of empty/null set(Φ) and universal set(U), $\Phi' = U$ and $U' = \Phi$.

Cartesian Product of Sets

If set A and set B are two sets then the cartesian product of set A and set B is a set of all ordered pairs (a, b), such that a is an element of A and b is an element of B. It is denoted by A × B. We can represent it in set-builder form, such as:

$$A \times B = \{(a, b) : a \in A \text{ and } b \in B\}.$$

Example: set A = {1, 2, 3} and set B = {Bat, Ball}, then;

$$A \times B = \{(1, \text{Bat}),(1, \text{Ball}),(2, \text{Bat}),(2, \text{Ball}),(3, \text{Bat}),(3, \text{Ball})\}.$$

Difference of Sets

If set A and set B are two sets, then set A difference set B is a set which has elements of A but no elements of B. It is denoted as A – B.

Example: A = {1, 2, 3} and B = {2, 3, 4};

$$A - B = \{1\}.$$

Sets Formulas

Some of the most important set formulas are:

For any three sets A, B and C,

$$n(A \cup B) = n(A) + n(B) - n(A \cap B)$$

If $A \cap B = \varnothing$, then $n(A \cup B) = n(A) + n(B)$

$$n(A - B) + n(A \cap B) = n(A)$$

$$n(B - A) + n(A \cap B) = n(B)$$

$$n(A - B) + n(A \cap B) + n(B - A) = n(A \cup B)$$

$$n(A \cup B \cup C) = n(A) + n(B) + n(C) - n(A \cap B) - n(B \cap C) - n(C \cap A) + n(A \cap B \cap C)$$

Properties of Sets

Commutative Property:

- $A \cup B = B \cup A$
- $A \cap B = B \cap A$

Associative Property:

- $A \cup (B \cup C) = (A \cup B) \cup C$
- $A \cap (B \cap C) = (A \cap B) \cap C$

Distributive Property:

- $A \cup (B \cap C) = (A \cup B) \cap (A \cup C)$
- $A \cap (B \cup C) = (A \cap B) \cup (A \cap C)$

De Morgan's Law:

- Law of union : $(A \cup B)' = A' \cap B'$
- Law of intersection : $(A \cap B)' = A' \cup B'$

Complement Law:

- $A \cup A' = A' \cup A = U$
- $A \cap A' = \varnothing$

Idempotent Law and Law of a null and universal set:

For any finite set A:

- $A \cup A = A$
- $A \cap A = A$
- $\varnothing' = U$
- $\varnothing = U'$

Binary Operation

Binary operation is an operation that requires two inputs. These inputs are known as operands. The binary operation of addition, multiplication, subtraction and division takes place on two operands. Even when we add any three binary numbers, we first add two numbers and then the third number will be added to the result of the two numbers. Thus, the mathematical operations which are done with the two numbers are known as binary operations.

The binary operation conjoins any two elements of a set. The results of the operation of binary numbers belong to the same set. Let us take the set of numbers as X on which binary operations will be performed. Now, we will perform binary operations such as addition, subtraction, multiplication and division of two sets (a and b) from the set X. The result of the operation (a and b) will be the other element that will belong to the same set X.

Hence, the binary operation is stated as an operation which is performed on set X. This function is derived by * A * A. Thus, the binary operation * performed on operands a and b is symbolized as a*b.

Binary Operation Examples

Let us understand the binary addition on natural numbers and real numbers. If we add two operands which are natural numbers such as x and y, the result of this operation will also be a natural number. Same rule holds for real numbers as well.

$+: R + R \rightarrow R$ is derived by $(x, y) \rightarrow x + y$

$+: N + N \rightarrow N$ is derived by $(x, y) \rightarrow x + y$.

Let us understand the binary multiplication on natural numbers and real numbers. If we multiply two operands which are natural numbers such as x and y, the result of this operation will also be a natural number. Same rule holds for real numbers as well.

$+: R \times R \rightarrow R$ is derived by $(x, y) \rightarrow x \times y$

$+: N \times N \rightarrow N$ is derived by $(x, y) \rightarrow x \times y$.

Let us understand the binary subtraction on natural numbers and real numbers. If we subtract two operands which are real numbers such as x and y, the result of this operation will also be a natural number. Same rule does not hold for natural numbers because if we take two numbers such as x and y and perform binary subtraction on it, then the result will not be in real numbers.

For example = 3 - 4 = -1 (-1 is not a real number).

Hence,

$-: R - R \rightarrow R$ is derived by $(x\ y) \rightarrow x - y$.

Let us understand the binary division on natural numbers and real numbers. If we divide two operands which are real numbers such as x and y, the result of this operation will also be a natural number. Same rule does not hold for natural numbers because if we take two numbers such as x and y and perform binary division on it, then the result will not be in real numbers.

For example: $1 \div 0 = 0$ (0 is not a real number),

Hence, $\div : R \div R \rightarrow R$ is derived by $(x, y) \rightarrow x \div y$.

Binary Operation Types

Binary operations such as binary addition, binary subtraction, binary multiplication and binary division are calculated similarly as the arithmetic operations are calculated in numerals.

There are four types of binary operations namely:

Binary Addition

The result obtained after adding two binary numbers is the binary number itself. Binary addition is the simplest method to add any of the binary numbers. It can be calculated easily if we know the following rules:

Rules:

- $0 + 0 = 0$
- $0 + 1 = 1$
- $1 + 0 = 1$
- $1 + 1 = 10$

Let us take any two binary numbers and add them.

Add : $10001 + 11101 = 101110$

Binary Subtraction

The result obtained after subtracting two binary numbers is the binary number itself. Binary subtraction is also the simplest method to subtract any of the binary numbers. It can be calculated easily if we know the following rules:

Rules:

- $0 - 0 = 0$
- $0 - 1 = 1$ (with a borrow of 1)
- $1 - 0 = 1$
- $1 - 1 = 0$

Let us take any two binary numbers and subtract them.

Binary Multiplication

The binary multiplications are calculated similarly as the other arithmetics numerals are calculated. Let us take any two binary numbers and multiply them. It can be calculated easily if we know the following rules:

Rules:

- $0 \times 0 = 0$
- $0 \times 1 = 0$
- $1 \times 0 = 0$
- $1 \times 1 = 1$

Example: 1101 * 1010 = 10000010

Binary Division

The method of binary division is similar to the 10 decimal system other than the base 2 system. It can be calculated easily if we know the following rules:

- $1 \div 1 = 1$
- $1 \div 0 = 0$
- $0 \div 1 =$ Meaningless
- $0 \div 0 =$ Meaningless

Identity Element

An identity element in a set is an element that is special with respect to a binary operation on the set: when an identity element is paired with any element via the operation, it returns that element.

More explicitly, let S be a set and $*$ be a binary operation on S. Then,

- An element $e \in S$ is a left identity if $e * s = s$ for any $s \in S$.
- An element $f \in S$ is a right identity if $s * f = s$ for any $s \in S$.
- An element that is both a left and right identity is called a two-sided identity, or identity element, or identity for short.

Example: Let $S = \{a, b, c, d\}$, and consider the binary operation defined by the following table:

*	a	b	c	d
a	a	a	a	a
b	c	b	d	b
c	d	c	b	c
d	a	b	c	d

The value of $x * y$ is given by looking up the row with x and the column with y.

What are the Left Identities? What are the Right Identities?

The unique left identity is d. The unique right identity is also d. This is because the row corresponding to a left identity should read a, b, c, d, as should the column corresponding to a right identity. Note that $*$ is not a commutative operation ($x * y$ and $y * x$ are not necessarily the same), so a left identity is not automatically a right identity (imagine the same table with the top right entry changed from a to something else).

Example: Let, $S = \mathbb{R}$, the set of real numbers, and let $*$ be addition. Then 0 is an identity element: $0 + s = s + 0 = s$ for any $s \in \mathbb{R}$. Note that 0 is the unique left identity, right identity, and identity element in this case.

Existence and Properties

In general, there may be more than one left identity or right identity; there also might be none.

Let S = ℝ, and define $*$ by the formula,

$$a * b = a^2 - 3a + 2 + b.$$

What are the Left Identities, Right Identities and Identity Elements?

If e is a left identity, then $e * b = b$ for all $b \in \mathbb{R}$ so $e^2 - 3e + 2 + b = b$, so $e^2 - 3e + 2 = 0$. This has two solutions, e = 1, 2, so 1 and 2 are both left identities.

If f is a right identity, then $a * f = a$ for all $a \in \mathbb{R}$, so $a = a^2 - 3a + 2 + f$, so $f = -a^2 + 4a - 2$. But no f can be equal to $-a^2 + 4a - 2$ for all $a \in \mathbb{R}$: for instance, taking a = 0 gives $f = -2$, but taking a = 1 gives f = 1. This is impossible. So there are no right identities. Because there is no element which is both a left and right identity, there is no identity element.

It is the case that if an identity element exists, it is unique: If S is a set with a binary operation, and e is a left identity and f is a right identity, then e = f and there is a unique left identity, right identity, and identity element.

Proof: By the properties of identities,

$$e = e * f = f.$$

If e' is another left identity, then e' = f by the same argument, so e' = e. So the left identity is unique. A similar argument shows that the right identity is unique. Since e = f, it is both a left and a right identity, so it is an identity element, and any other identity element must equal it, by the same argument.

Inverse Elements

Given an element a in a set with a binary operation, an inverse element for a is an element which gives the identity when composed with a. More explicitly, let S be a set, $*$ a binary operation on S, and $a \in S$. Suppose that there is an identity element e for the operation. Then,

- An element b is a left inverse for a if $b * a = e$;
- An element c is a right inverse for a if $a * c = e$;
- An element is an inverse (or two-sided inverse) for a if it is both a left and right inverse for a.

Example: Let S = {a, b, c, d}, and consider the binary operation defined by the following table:

*	a	b	c	d
a	a	a	a	a
b	c	b	d	b
c	d	c	b	c
d	a	b	c	d

The value of $x * y$ is given by looking up the row with x and the column with y.

Which elements have left inverses and right inverses?

Since d is the identity and $b * c = c * a = d * d = d$, it follows that:

- a has no left or right inverse.
- b has no left inverse, and has a right inverse c.
- c has a left inverse b, and a right inverse a.
- d is its own left and right inverses.

Example: Consider the set $\mathbb{R}$ with the binary operation of addition. The identity element is 0, so the inverse of any element a is $-a$, as $(-a)+a = a+(-a) = 0$. So every element has a unique left inverse, right inverse, and inverse.

Existence and Properties of Inverse Elements

An element might have no left or right inverse, or it might have different left and right inverses, or it might have more than one of each.

Example: Let S be the set of functions $f : \mathbb{R} \to \mathbb{R}$. There is a binary operation given by composition $f * g = f \circ g$, i.e., $(f * g)(x) = f(g(x))$. The (two-sided) identity is the identity function $i(x) = x$.

Find a function with more than one left inverse. Find a function with more than one right inverse.

If $f(x) = e^x$, then f has more than one left inverse: let,

$$g_1(x) = \begin{cases} \ln(|x|) & \text{if } x \neq 0 \\ 0 & \text{if } x = 0 \end{cases},$$

and let,

$$g_2(x) = \begin{cases} \ln(|x|) & \text{if } x > 0 \\ 0 & \text{if } x \leq 0 \end{cases}.$$

Then $g_1(f(x)) = \ln(|e^x|) = \ln(e^x) = x$, and $g_2(f(x)) = \ln(e^x) = x$, because e^x is always positive. The idea is that g_1 and g_2 are the same on positive values, which are in the range of f, but differ on negative values, which are not.

If,

$$f(x) = \begin{cases} \tan(x) & \text{if } \sin(x) \neq 0 \\ 0 & \text{if } \sin(x) = 0 \end{cases},$$

then f has more than one right inverse: let $g_1(x) = \arctan(x)$ and $g_2(x) = 2\pi + \arctan(x)$. Then $f(g_1(x)) = f(g_2(x)) = x$.

The first example was injective but not surjective, and the second example was surjective but not injective.

If the binary operation is associative and has an identity, then left inverses and right inverses coincide:

Theorem: If S is a set with an associative binary operation $*$ with an identity element, and an element $a \in S$ has a left inverse b and a right inverse c, then b = c and a has a unique left, right, and two-sided inverse.

Proof: Let e be the identity. Then,

$$c = e * c = (b * a) * c = b * (a * c) = b * e = b.$$

The same argument shows that any other left inverse b' must equal c, and hence b. Similarly, any other right inverse equals b, and hence c. So there is exactly one left inverse and exactly one right inverse, and they coincide, so there is exactly one two-sided inverse.

Examples of Inverse Elements

The existence of inverses is an important question for most binary operations. Here are some examples:

- Let S = $\mathbb{R}$ with $a * b = ab + a + b$. It is straightforward to check that this is an associative binary operation with two-sided identity 0. Then the inverse of a, if it exists, is the solution to $ab + a + b = 0$, which is $b = -\frac{a}{a+1}$, but when a = - 1 this inverse does not exist; indeed $(-1) * b = b * (-1) = -1$ for all b. So every element of $\mathbb{R}$ has a two-sided inverse, except for -1.
- Let $\mathbb{R}^\infty$ be the set of sequences $(a_1, a_2, a_3, \ldots)$ where the a_i are real numbers. Let S be the set of functions $f : \mathbb{R}^\infty \to \mathbb{R}^\infty$. Then composition of functions is an associative binary operation on S, with two-sided identity given by the identity function. Now let t be the shift operator, $t(a_1, a_2, a_3) = (0, a_1, a_2, a_3, \ldots)$. Then t has many left inverses but no right inverses (because t is injective but not surjective). One of its left inverses is the reverse shift operator $u(b_1, b_2, b_3, \ldots) = (b_2, b_3, \ldots)$.
- Let G be a group. Then every element of the group has a two-sided inverse, even if the group is nonabelian (i.e. the operation is not commutative).
- Let R be a ring. Then every element of R has a two-sided additive inverse (R is a group under addition), but not every element of R has a multiplicative inverse. In particular, 0_R never has a multiplicative inverse, because $0 \cdot r = r \cdot 0 = 0$ for all $r \in R$. If every other element has a multiplicative inverse, then R is called a division ring, and if R is also commutative, then it is called a field. In general, the set of elements of R with two-sided multiplicative inverses is called R*, the group of units of R.

Group Theory

The group theory is the most crucial ingredient in the present day of science, mathematics, statistics and computer science. It was ascertained in the nineteenth century in association with delivering solutions for algebraic expressions.

In particular, the group was the set of all the permutations of the roots of an algebraic expression that exhibits the characteristics that the combination of any two of these permutations belongs to the set. And later on, the belief was made generalized to the notion of an abstract group. However, an abstract group is the study of a set, with an operation defined on it.

There are four primary sources in the development of group theory, which are (in the terms of author names and time of origin):

- Classical algebra,
- Number theory,
- Geometry,
- Analysis.

The theory of group is essentially the study of groups where a group is a set equipped with specific binary operations. For example, the set of all the integers with addition. If there are a finite number of components, the group is termed as a finite group, where the number of components is called the group order of the group. A subset of the group which is bounded under any of the group operations is known as Subgroups.

More specifically, the formal definition of the group is:

- The group involves identity.
- The group includes inverse.
- The group holds an operation that is associative.
- The group is restricted under an operation.

In laymen terms, group theory is the study of the set of components present in a group, where a group is the acquisition of the components/elements that are integrated together to perform some operations on them.

The group theory is the branch of abstract-algebra that is incurred for studying and manipulating abstract concepts involving symmetry. It is the tool which is used to determine the symmetry. Also, symmetry operations and symmetry components are two fundamental and influential concepts in group theory. In core words, group theory is the study of symmetry, therefore while dealing with the object that exhibits symmetry or appears symmetric, group theory can be used for analysis.

The simple answer is that Group Theory is the systematic and methodical study of symmetry, for an instance, when a physical quantity possess some sort of symmetry and so it undergoes any kind of symmetry operation in order to obtain a simplified description through accounting the outcomes of that symmetry.

Outcomes, obtained from group theory, could only be useful if and only if an individual understands them well enough to consider and deploy them up and provide users with a few basics insights as earlier as possible. Group theory appears in the situation where the symmetry plays a significant part and its roots fall in Galois' study of the symmetries of the roots of polynomials.

Properties of Group under Group Theory

A group, G, is a finite or infinite set of components/factors, unitedly through a binary operation or group operation, that jointly meet the four primary properties of the group, i.e closure, associativity, the identity, and the inverse property.

"Group theory is the natural language to describe the symmetries of a physical system."

The operation (or formula) by virtue of which a group is determined is known as "Group Operation" and a set is treated as a group under that operation. Consider the elements A, B, C ....with binary operations amid A and B, and then notation "AB" form a group if:

- Closure: If A and B are two components in the group, G, then the product "AB" is also in "G".
- Associativity: The defined multiplication is associative, therefore, for all A, B, C are in the group, "G", then (AB) C= A (BC).
- Identity: For any component, A, there also exists the identity element, I, such that IA= AI= A.
- Inverse: There should be an inverse of each component, so, for every component A under G, the set incorporates a component B= A' such that AA'= A'A= I.

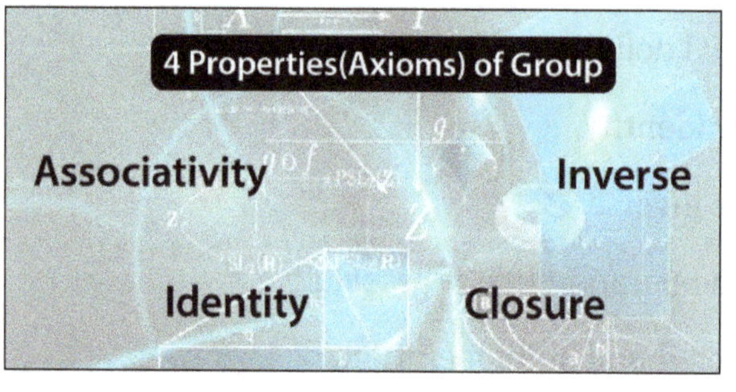

Some other fundamental properties include:

- A group is a monoid, where each of its components is invertible.
- A group must comprise at least a single component, with the unique isomorphism, where the single-element group recognised as the trivial group.
- The group is abelian if, for any A, B are elements of G, then AB = BA.

What are the General Application of Group Theory?

Group theory has multiple facets and various beneficial applications within and outside a broad width of science and mathematics, also groups mount in a number of supposedly impertinent entities, for example, they resemble in crystallography and quantum mechanics, in geometry and topology, in analysis and algebra, etc.

- In Physics, it plays an indispensable role in the determination of the selection rules for spectroscopic transitions in Atomic and Molecular Spectroscopy.
- In Chemistry, it is implemented to study and analyze the symmetries and the crystal

structures of molecules, including many physical and chemical properties, and spectroscopic properties of the molecule. The group theory has turned out to be a standard and a powerful tool for studying molecular properties in the terms of molecular orbital theory.

- Group theory, being an eventual and powerful tool for symmetry, has an ultimate impact on research in robotics, computer vision, computer graphics and medical image analysis.
- In Mathematics, it can be used for classification of identical mathematical objects that possess symmetry, for example in geometric figures (a circle is highly symmetric and invariant under any rotation) and in mathematical functions and operations.
- For the smooth data transmission, the concept of group, subgroups and cosets are used in Cryptography and Public Key algorithms.
- In recent researches, it has exploited in music screening as well.

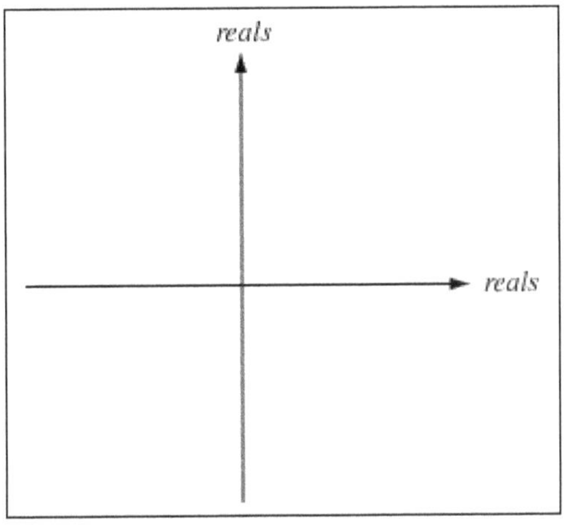

Given two groups G and H, there are several ways to form a new group. The simplest is the direct product, denoted G × H. As a set, the group direct product is the Cartesian product of ordered pairs (g, h), and the group operation is component wise, so,

$$(g_1, h_1) \times (g_2, h_2) = (g_1 g_2, h_1 h_2).$$

For example, $\mathbb{R} \times \mathbb{R}$ is isomorphic to $\mathbb{R}^2$ under vector addition. In a similar fashion, one can take the direct product of any number of groups by taking the Cartesian product and operating component wise.

G is isomorphic to the subgroup of elements (g, e_H) where e_H is the identity element in H. Similarly, H can be realized as a subgroup. The intersection of these two subgroups is the identity (e_G, e_H), and the two subgroups are normal.

$$\begin{array}{c} X \to G \\ \downarrow \\ H \end{array} \quad \Rightarrow \quad \begin{array}{c} X \to G \oplus H \to G \\ \downarrow \\ H \end{array}$$

The universal property of a direct product; X factors through $G \oplus H$.

Like the ring direct product, the group direct product has the universal property that if any group X has a homomorphism to G and a homomorphism to H, then these homomorphisms factor through $G \times H$ in a unique way.

If one has group representations R_G of G and R_H of H, then there is a representation $R_G \otimes R_H$ sometimes called the external tensor product, given by the ⊗ product ⊗. In this case, the group character satisfies,

$$\mathcal{X}(g \otimes h) = \mathcal{X}_{R_G}(g)\, \mathcal{X}_{R_H}(h).$$

Direct Product of Group

External

We define a suitable binary operation $*$ on the cartesian product $G_1 \times G_2$ of two groups G_1 and G_2 so that $(G_1 \times G_2, *)$ is a group which we call the external direct product of groups. This is a widely used technique to construct a new group from old as well as to decompose a group as a product of relatively better known groups. Thus external direct product of groups is an important notion in the structure theory of finite groups.

Let $G_1, G_2, \cdots, G_n$ be n groups. Define a binary operation $*$ on the cartesian product $G = G_1 \times G_2 \times \cdots \times G_n$ by:

$$(a_1, a_2, \cdots, a_n) * (b_1, b_2, \cdots, b_n) = (a_1 b_1, a_2 b_2, \cdots, a_n b_n)$$

for every $(a_1, a_2, \cdots, a_n), (b_1, b_2, \cdots, b_n) \in G$. Here $a_i b_i$ is the product of two elements a_i and b_i in G_i for every $i = 1, 2, \cdots, n$.

For example, the binary operation $*$ on $K_4 \times \mathbb{Z}$ is given by:

$$(a, m) * (b, n) = (ab, m + n), \text{ for every } (a, m), (b, n) \in K_4 \times \mathbb{Z}.$$

It is easy to check that $(K_4 \times \mathbb{Z}, *)$ is a group having the identity element (e, 0) and for every (a, n) $\in K_4 \times \mathbb{Z}$, $(a, n)^{-1} = (a, -n)$.

Similarly (G, $*$) is a group with the identity element $e = (e_1, e_2, \cdots, e_n)$ and $(a_1, a_2, \cdots, a_n)^{-1} = (a_1^{-1}, a_2^{-1}, \cdots, a_n^{-1})$ for every $(a_1, a_2, \cdots, a_n) \in G$ where e_i is the identity element of G_i for every $i = 1, 2, \cdots, n$.

Let $G_1, G_2, \cdots, G_n$ be n groups and $G = G_1 \times G_2 \times \cdots \times G_n$. Then the group (G, $*$) is called the external direct product of the groups $G_1, G_2, \cdots, G_n$.

Example: Consider the noncommutative group (S_3) and the infinite group $(\mathbb{Z}, +)$. Then the binary operation $*$ on $G = S_3 \times \mathbb{Z}$ is given by:

$$(\alpha, m) * (\beta, n) = (\alpha\beta, m + n).$$

The identity element in G is (e, 0) and $(\alpha, m)^{-1} = (\alpha^{-1}, -m)$ for every $\alpha \in S_3$ and $m \in \mathbb{Z}$. For example,

$$((12),\ 4)^{-1} = ((12), -4)$$
$$\text{and } ((123),\ 2)^{-1} = ((132), -2).$$

Now,

$$((12),\ 4) * ((13),\ 7) = ((12)(13),\ 4+7) = ((132), 11)$$
$$\text{and } ((13), 7) * ((12),\ 4) = ((13)(12),\ 7+4) = ((123), 11).$$

shows that the group $S_3 \times \mathbb{Z}$ is noncommutative. Also $S_3 \times \mathbb{Z}$ is an infinite group. What makes it interesting is that noncommutativity of $S_3 \times \mathbb{Z}$ comes from S_3, whereas infiniteness of $\mathbb{Z}$ makes $S_3 \times \mathbb{Z}$ an infinite group.

Theorem: Let G_1 and G_2 be two groups. Then $G_1 \times G_2 \simeq G_2 \times G_1$.

Proof: Define $f : G_1 \times G_2 \to G_2 \times G_1$ by:

$$f(a,\ b) = (b,\ a) \text{ for all } (a,\ b) \in G_1 \times G_2.$$

Then,

$$\begin{aligned} f((a_1,\ b_1) * (a_2,\ b_2)) &= f(a_1a_2,\ b_1b_2) \\ &= (b_1b_2,\ a_1a_2) \\ &= (b_1,\ a_1) * (b_2,\ a_2) \\ &= f(a_1,\ b_1) * f(a_2,\ b_2). \end{aligned}$$

shows that f is a homomorphism. Also f is one-to-one and onto. Thus f is an isomorphism, and hence $G_1 \times G_2 \simeq G_2 \times G_1$.

Theorem: Let $G_1, G_2, \cdots, G_n$ be n-groups. Then the group $G = G_1 \times G_2 \times \cdots \times G_n$ is abelian if and only if each of the groups $G_1, G_2, \cdots, G_n$ is abelian.

Proof: First assume that G is abelian. Consider $1 \le i \le n$ and $a, b \in G_i$. Then $(e_1, \cdots, e_{i-1}, a, e_{i+1}, \cdots, e_n)$, $(e_1, \cdots, e_{i-1}, b, e_{i+1}, \cdots, e_n) \in G$ and commutativity of G implies that,

$$\begin{aligned} &(e_1, \cdots, e_{i-1},\ a,\ e_{i+1}, \cdots, e_n) * (e_1, \cdots, e_{i-1},\ b,\ e_{i+1}, \cdots, e_n) \\ &\qquad = (e_1, \cdots, e_{i-1},\ b,\ e_{i+1}, \cdots, en) * (e_1, \cdots, e_{i-1},\ a,\ e_{i+1}, \cdots, e_n) \\ &\Rightarrow (e_1, \cdots, e_{i-1},\ ab,\ e_{i+1}, \cdots, e_n) = (e_1, \cdots, e_{i-1},\ ba,\ e_{i+1}, \cdots, e_n) \\ &\Rightarrow ab = ba. \end{aligned}$$

Thus G_i is abelian for every $1 \le i \le n$.

Conversely, suppose that G_i is abelian for every $1 \le i \le n$. Consider $(a_1, a_2, \cdots, a_n), (b_1, b_2, \cdots, b_n) \in G$. Then $a_i, b_i \in G_i$, and since G_i is abelian, so $a_ib_i = b_ia_i$ for every $1 \le i \le n$. Now,

$$\begin{aligned}(a_1, a_2, \cdots, a_n) * (b_1, b_2, \cdots, b_n) &= (a_1b_1, a_2b_2, \cdots, a_nb_n)\\ &= (b_1a_1, b_2a_2, \cdots, b_na_n)\\ &= (b_1, b_2, \cdots, b_n) * (a_1, a_2, \cdots, a_n).\end{aligned}$$

implies that G is abelian. Following generalization of this theorem shows that even if G_1 and G_2 are not abelian then also the extent of commutativity can be measured by that of G_1 and G_2.

Theorem: Let $G_1, G_2, \cdots, G_n$ be n-groups. Then $Z(G_1 \times G_2 \times \cdots \times G_n) \simeq Z(G_1) \times Z(G_2) \times \cdots \times Z(G_n)$.

Proof: This is sufficient to prove the result for n = 2. Let G_1 and G_2 be two groups. Consider $a \in G_1$ and $b \in G_2$. Then we have,

$$\begin{aligned}&(a, b) \in Z(G_1 \times G_2)\\ &\Leftrightarrow (a, b)*(g_1, g_2) = (g_1, g_2)*(a, b) \text{ for all } (g_1, g_2) \in G_1 \times G_2\\ &\Leftrightarrow (ag_1, bg_2) = (g_1a, g_2b) \text{ for all } (g_1, g_2) \in G_1 \times G_2\\ &\Leftrightarrow ag_1 = g_1a \text{ and } bg_2 = g_2b \text{ for all } g_1 \in G_1,\ g_2 \in G_2\\ &\Leftrightarrow a \in Z(G_1) \text{ and } b \in Z(G_2)\\ &\Leftrightarrow (a, b) \in Z(G_1) \times Z(G_2).\end{aligned}$$

Thus $Z(G_1 \times G_2) \simeq Z(G_1) \times Z(G_2)$.

Now we show that the product of two cyclic groups may not be cyclic. For example, consider the external direct product $\mathbb{Z}_4 \times \mathbb{Z}_6$ of the cyclic groups $\mathbb{Z}_4$ and $\mathbb{Z}_6$. Then $\mathbb{Z}_4 \times \mathbb{Z}_6$ contains 24 elements, but, for every $([a]_4, [b]_6) \in \mathbb{Z}_4 \times \mathbb{Z}_6$, $12([a]_4, [b]_6) = ([12a]_4, [12b]_6) = ([0]_4, [0]_6)$ implies that $\mathbb{Z}_4 \times \mathbb{Z}_6$ has no element of order 24. Therefore $\mathbb{Z}_4 \times \mathbb{Z}_6$ is not a cyclic group. Also this example gives us a hope that the group $\mathbb{Z}_m \times \mathbb{Z}_n$ may be cyclic if lcm (m, n) = mn, equivalently, if m and n are relatively prime.

Now we prove a necessary and sufficient condition for the external direct product of two finite cyclic groups to be cyclic.

Theorem: Let G_1 and G_2 be two finite cyclic groups of order m and n respectively. Then $G_1 \times G_2$ is cyclic if and only if gcd(m, n) = 1.

In this case, (a, b) is a generator of $G_1 \times G_2$ if and only if a and b are generators of G_1 and G_2 respectively.

Proof: First assume that $G_1 \times G_2$ is a cyclic group and $G_1 \times G_2 = \langle (a, b) \rangle$. Then o((a, b)) = mn. Let d = gcd(m, n). Now $(a,b)\frac{mn}{d} = \left(a^{m\frac{n}{d}}, b^{n\frac{m}{d}}\right) = (e_1, e_2)$ implies that $mn \mid \frac{mn}{d}$. Hence it follows that d = 1.

Conversely, let gcd(m, n) = 1. Suppose that $G_1 = \langle a \rangle$ and $G_2 = \langle b \rangle$. Denote k = o(a, b). Then

$(a, b)^{mn} = (a^{mn}, b^{mn}) = (e_1, e_2)$ implies that $k \mid mn$. Now $(a^k, b^k) = (a, b)^k = (e_1, e_2)$ implies that $a^k = e_1$ and $b^k = e_2$; and so $m \mid k$ and $n \mid k$. Then $\gcd(m, n) = 1$ implies that $mn \mid k$. Therefore $k = mn = |G_1 \times G_2|$. Hence $G_1 \times G_2$ is a cyclic group and (a, b) is a generator.

The following consequence is immediate:

Corollary: $\mathbb{Z}_m \times \mathbb{Z}_n \simeq \mathbb{Z}_{mn}$ if and only if $\gcd(m, n) = 1$.

Thus we see that for $a \in G_1$ of order m and $b \in G_2$ of order n, if $\gcd(m, n) = 1$, then $o((a, b)) = mn$. What does it happen if $\gcd(m, n) \neq 1$?

Theorem: Let G_1 and G_2 be two finite groups. Then for every $a \in G_1$, $b \in G_2$, the order of (a, b) in $G_1 \times G_2$ is $o((a, b)) = \text{lcm}(o(a), o(b))$.

Proof: Since G_1 and G_2 are finite groups, so $G_1 \times G_2$ is finite. Thus for every $a \in G_1$, $b \in G_2$, o(a) in G_1, o(b) in G_2 and o((a, b)) in $G_1 \times G_2$ all are finite. Now we have,

$$
\begin{aligned}
& o(a,b) \text{ is the smallest positive integer n such that } (a,b)^n = (e_1, e_2), \\
\Leftrightarrow\ & \text{n is the smallest positive integer such that } a^n = e_1 \text{ and } b^n = e_2, \\
\Leftrightarrow\ & \text{n is the smallest positive integer such that } o(a) \mid n \text{ and } o(b) \mid n.
\end{aligned}
$$

Thus o(a, b) = lcm (o(a), o(b)).

External direct product of two infinite cyclic groups is not cyclic. Since every infinite cyclic group is isomorphic to $\mathbb{Z}$, it is sufficient to consider $\mathbb{Z} \times \mathbb{Z}$. On the contrary, if possible, let $\mathbb{Z} \times \mathbb{Z}$ be cyclic and $\mathbb{Z} \times \mathbb{Z} = \langle (a, b) \rangle$. Since $(1, 1) \in \mathbb{Z} \times \mathbb{Z}$, so there is $r \in \mathbb{Z}$ such that $r(a, b) = (1, 1)$ which implies that $a = b = 1$ or -1. Then $\langle (a, b) \rangle = \{(n, n) \mid n \in \mathbb{Z}\} \neq \mathbb{Z} \times \mathbb{Z}$. This contradiction shows that $\mathbb{Z} \times \mathbb{Z}$ is not cyclic.

We conclude this module with the characterization of quotients of $G_1 \times G_2$.

Theorem: Let G_1 and G_2 be two groups. Then,

$$G_1 \times G_2 / G_1 \times \{e_2\} \simeq G_2 \text{ and } G_1 \times G_2 / \{e_1\} \times G_2 \simeq G_1.$$

Proof: Define $f : G_1 \times G_2 \to G_2$ by:

$$f(a,b) = b \text{ for all } (a,b) \in H \times K.$$

Then f is a homomorphism and,

$$
\begin{aligned}
\ker f &= \{(a,b) \in G_1 \times G_2 \mid f(a,b) = e_2\} \\
&= \{(a,b) \in G_1 \times G_2 \mid b = e_2\} \\
&= \{(a,e_2) \mid a \in G_1\} \\
&= G_1 \times \{e_2\}.
\end{aligned}
$$

Also f is onto. Hence, by the First Isomorphism Theorem, $G_1 \times G_2/G_1 \times \{e_2\} \simeq G_2$. Similarly, we have $G_1 \times G_2/\{e_1\} \times G_2 \simeq G_1$.

Also we have:

Theorem: Let G_1 and G_2 be two groups. If N_1 and N_2 are two normal subgroups of G_1 and G_2 respectively, then $N_1 \times N_2$ is a normal subgroup of $G_1 \times G_2$, and

$$G_1 \times G_2 / N_1 \times N_2 \simeq G_1 / N_1 \times G_2 / N_2.$$

Proof: Define $f : G_1 \times G_2 \to G_1 / N_1 \times G_2 / N_2$ by:

$$f(a, b) = (aN_1, bN_2) \text{ for every } (a, b) \in G_1 \times G_2.$$

Then for every $(a, b),(c, d) \in G_1 \times G_2$,

$$\begin{aligned} f((a, b) * (c, d)) &= f(ac, bd) \\ &= (acN_1, bdN_2) \\ &= (aN_1cN_1, bN_2dN_2) \\ &= (aN_1, bN_2) * (cN_1, dN_2) \\ &= f(a, b) * f(c, d) \end{aligned}$$

implies that f is a homomorphism. Also f is onto. Then it follows from the First Isomorphism Theorem that,

$$G_1 \times G_2 / \ker f \simeq G_1 / N_1 \times G_2 / N_2.$$

Now we have,

$$\begin{aligned} \ker f &= \{(a, b) \in G_1 \times G_2 \mid f(a, b) = (N_1, N_2)\} \\ &= \{(a, b) \in G_1 \times G_2 \mid (aN_1, bN_2) = (N_1, N_2)\} \\ &= \{(a, b) \in G_1 \times G_2 \mid aN_1 = N_1 \text{ and } bN_2 = N_2\} \\ &= \{(a, b) \in G_1 \times G_2 \mid a \in N_1 \text{ and } b \in N_2\} \\ &= N_1 \times N_2. \end{aligned}$$

Thus the result follows.

Internal

Let $H_i \lhd G$ for $i \in \{1, \ldots, n\}$. [We require that H_i is normal!] Then G is the internal direct product of the H_i's if for any $g \in G$, $\exists! h_i \in H_i$ such that $g = h_1 \cdot h_2 \cdots h_n$. Here are the properties we gave to decide if a group is isomorphic to the (external) direct product of a finite number of its subgroups.

Let $H_i \leq G$ for $i \in \{1, \ldots, n\}$. Then the sets H_i satisfy the IDP properties if:

1. $H_i \lhd G$, for all i.

2. $G = H_1 \cdots H_n \stackrel{\text{def}}{=} \{h_1 \cdots h_n : h_i \in H_i\}$.
3. If $\hat{H}_i \stackrel{\text{def}}{=} H_1 \cdots H_{i-1} \cdot H_{i+1} \cdots H_n$, then $H_i \cap \hat{H}_i = \{1\}$. [If n = 2, then $\hat{H}_1 = H_2$ and $\hat{H}_2 = H_1$].

We will prove that the definitions are equivalent, i.e., G is the internal direct product of the H_i's if and only if the H_i's satisfy the IDP properties. We need the following lemma.

Lemma: If $H_i \leq G$ for $i \in \{1, \ldots, n\}$ satisfy IDP properties, then $h_i h_j = h_j h_i$ for all $h_i \in H_i$ and $h_j \in H_j$ with $i \neq j$.

Proof: Since $h_i^{-1} \in H_i \lhd G$, we have that $h_j h_i^{-1} h_j^{-1} \in H_i$. So $h_i(h_j h_i^{-1} h_j^{-1}) \in H_i$. Similarly, since $h_j \in H_j \lhd G$, we have that $h_i h_j h_i^{-1} \in H_j$. So $(h_i h_j h_i^{-1}) h_j^{-1} \in H_j$. Thus, we have that $h_i h_j h_i^{-1} h_j^{-1} \in H_i \cap H_j$. But since $i \neq j$, we have that $H_j \subset \hat{H}_i$. and so $H_i \cap H_j \subseteq H_i \cap \hat{H}_i$, Moreover, by property (3), we have that $H_i \cap \hat{H}_i = \{1\}$. Hence, $h_i h_j h_i^{-1} h_j^{-1} \in H_i \cap H_j \subseteq H_i \cap \hat{H}_i = \{1\}$, which implies that $h_i h_j h_i^{-1} h_j^{-1} = 1$, i.e., $h_i h_j = h_j h_i$.

We then have:

Theorem: Let $H_1, \ldots, H_n \leq G$. Then, $\varphi : H_1 \times \cdots \times H_n \to G$ defined by $\varphi(h_1, \ldots, h_n) = h_1 \cdots h_n$ is an isomorphism if and only if the H_i's satisfy the IDP properties.

Proof: [⇒:] Assume that φ [as in the statement] is an isomorphism. Let $\tilde{G} \stackrel{\text{def}}{=} H_1 \times H_n$ and $\tilde{H}_i \stackrel{\text{def}}{=} \{1\} \times \cdots \{1\} \times H_i \times \{1\} \times \cdots \times \{1\} \leq \tilde{G}$ [with H_i in the i-th coordinate]. Then, clearly $\phi(\tilde{H}_i) = H_i$. Since $\tilde{H}_i \lhd \tilde{G}$ [easy exercise!], we have that $H_i \lhd G$, as φ is an isomorphism [by assumption]. [This was a problem in the exam.] Thus, IDP property (1) is proved.

Since φ is an isomorphism [and hence onto] and $\varphi(\tilde{G}) = H_1 \cdots H_n$ [by definition of φ and the product of groups], we have that $G = H_1 \cdots H_n$, proving property (2).

Now, let $\hat{\hat{H}}_i \stackrel{\text{def}}{=} H_1 \times \cdots \times H_{i-1} \times \{1\} \times H_{i+1} \times \cdots \times H_n$. Then, clearly $\phi(\hat{\hat{H}}_i) = \hat{H}_i$ [with $\hat{H}_i$] and $\tilde{H}_i \cap \hat{\hat{H}}_i = \{(1, \ldots 1)\}$. Thus,

$$\begin{aligned}\{1\} &= \phi(\{(1, \ldots, 1)\}) \\ &= \phi\left(\tilde{H}_i \cap \hat{\hat{H}}_i\right) \\ &= \phi\left(\left(\tilde{H}_i\right) \cap \phi\, \hat{\hat{H}}_i\right) \\ &= H_i \cap \hat{H}_i\end{aligned}$$

Hence, property (3) is also satisfied.

[⇐:] Assume now that the H_i's satisfy the IDP property. Then, φ is a homomorphism by previous Lemma. It is onto by property (2) [as $\varphi(H_1 \times \cdots \times H_n) = H_1 \cdots H_n$ by definition of φ].

Now we show that φ is injective. Suppose that $\varphi(h_1, \ldots, h_n) = 1$. This means that $h_1 \cdots h_n = 1$, or $h_1^{-1} = h_2 \cdots h_n$. Since the left hand side is in H_1 and the right hand side is in $\hat{H}_1$, property (3) tells us that $h_1 = 1$ and $h_2 \cdots h_n = 1$. Then, $h_2^{-1} = h_3 \cdots h_n$ and now the left hand side is in H_2 and the right hand side is in $\hat{H}_2$ As before, we obtain $h_2 = 1$ and $h_3 \cdots h_n = 1$. Inductively, we obtain that $h_i = 1$ for all i. Hence, ker $\varphi = \{(1, \ldots, 1)\}$ and φ is injective.

Theorem: Let $H_i \lhd G$ for $i \in \{1, \ldots, n\}$. [We are already assuming that the H_i's are normal, since it is in the conditions of both definitions] We have that G is the internal direct product of the Hi if and only if the H_i's satisfy the IDP properties.

Proof: [⇒:] Assume that G is the internal direct product of the H_i's. Clearly properties (1) and (2) of IDP are satisfied.

Now, let $h_i \in H_i \cap \hat{H}_i$. Then, since $h_i \in \hat{H}_i$, we have, by definition, that,

$$1 \cdots 1 \cdot h_i \cdot 1 \cdots 1 = h_i = x_1 \cdots x_{i-1} \cdot 1 \cdot x_{i+1} \cdots x_n,$$

where $x_j \in H_j$. By the unique representation hypothesis, we have that $h_i = 1$. Thus $H_i \cap \hat{H}_i = \{1\}$,, i.e., property (3) is also satisfied.

[⇐:] Assume now that the H_i's satisfy the IDP properties. That the H_i's are normal, but we are already assuming it here.] Every element $g \in G$ can be written as $g = h_1 \cdots h_n$ with $h_i \in H_i$. [We need to show uniqueness.] Now assume that,

$$h_i \cdots h_n = x_1 \cdots x_n \quad \text{with}\, h_i, x_i \in H_i.$$

Thus, with φ as in the statement of previously described Theorem [which we can use since are assuming IDP properties], we have that,

$$\phi(h_1, \ldots, h_n) = \phi(x_1, \ldots, x_n).$$

Since φ is an isomorphism [and hence one-to-one], we have that $h_i = x_i$ for all i, and hence the representation is unique. This gives us:

Corollary: G is the internal direct product of the subgroups H_i's for $i \in \{1, \ldots, n\}$ [and hence $H_i \lhd G$ by assumption!] if and only if $\phi : H_1 \times \cdots \times H_n \rightarrow G$ defined by $\phi(h_1, \ldots, h_n) = h_1 \cdots h_n$ is an isomorphism.

Proof: By previously described Theorem, we know that that H_i's satisfying IDP is equivalent to φ [as in the statement] being an isomorphism. Since the former is equivalent to G being the internal direct product of the subgroups H_i's, the result follows.

Cauchy's Theorem

We have examples of finite noncommutative groups G such that the converse of the Lagrange's Theorem does not hold on G. Then it is natural either to proceed with the groups of order of particular form or to consider divisors of particular form, viz. a prime number or a power of a prime number. Cauchy's Theorem is one which proves that every finite group G contains an element and hence a subgroup of order p, if p is a prime divisor of $|G|$.

Lemma: Let G be a finite abelian group. If a prime $p \mid |G|$, then G has an element of order p.

Proof: Let G_p be the p-primary component of G. Then $|G_p| = p^n$ for some $n \geq 1$. Suppose that a is a nonzero element of G_p. Then $o(a) \mid |G_p|$ implies that $o(a) = p^m$. Since a is nonzero, so $m \geq 1$. Then $p^{m-1}a \in G_p$ is of order p. Therefore G has an element of order p.

Theorem: (Cauchy's Theorem) - Let G be a finite group. If a prime $p \mid |G|$, then G has an element of order p.

Proof: Let $|G| = p^n$. We prove this theorem by induction on n. If n = 1, then |G| = p. Hence not only one element but also every nonidentity element of G is of order p. Let us make our induction hypothesis that every group with p^k elements contains an element of order p for every $k < n$.

Now we prove the induction step. Consider the class equation,

$$|G| = |Z(G)| + \sum_{a \notin Z(G)} [G : C(a)]$$

of G. If G = Z(G), then G is abelian and the result follows by Lemma above. Suppose that $G \neq Z(G)$. For every $a \notin Z(G)$, $C(a) \subsetneq G$, and hence $|C(a)| < |G|$. If for some a $a \notin Z(G)$, $p \mid |C(a)|$ then, by the induction hypothesis, there is an element a of order p in C(a) and hence in G. Otherwise, for every $a \notin Z(G), p \nmid |c(a)|$. It follows from the Lagrange's Theorem that $p \mid |G| = [G : C(a)]|C(a)|$ and hence $p \mid [G : C(a)]$. Then $p \mid \sum_{a \notin Z(G)} [G : C(a)]$ and the class equation implies that $p \mid |Z(G)|$. It follows, by Lemma above, that Z(G) and hence G contains an element of order p.

So far, we have results on the possible order of the elements of a finite group G, when |G| is given. Cauchy's Theorem can be used to get an idea of the order of a finite group G, when we are given the orders of all the elements of G.

Theorem: Let G be a finite abelian group such that $|G| > 1$. Then G is simple if and only if G is a cyclic group of order p, for some prime p.

Proof: Let G be a simple group. Since $|G| > 1$, so there is a prime p such that $p \mid |G|$. Then, by the Cauchy's Theorem, G has an element, say a of order p. Denote $H = \langle a \rangle$. Then $p > 1$ implies that $H \neq \{e\}$. Since G is abelian, so H is normal in G. Then simplicity of G implies that $G = H = \langle a \rangle$. Thus G is a cyclic group and $|G| = o(a) = p$.

The converse follows directly from the Lagrange's Theorem. Now we show that the finite groups of order of the form p n where p is a prime and $n \geq 0$, posses many interesting features.

Let p be a prime. A group G is called a p-group if order of every element of G is p^n for some integer $n \geq 0$. A subgroup H of a group G is called a p-subgroup if H is a p-group.

Example: In $\mathbb{Z}_4$, order of $\bar{\ }$ is $1 = 2^0$, order of $\bar{2}$ is $2 = 2^1$ and order of $\bar{1}$ and $\bar{3}$ is $4 = 2^2$. Thus $\mathbb{Z}_4$ is a 2-group.

In S_3, there are elements of order 2 as well as of order 3. Since there is no prime p such that $2 = p^r$ and $3 = p^s$ for some $r, s \geq 0$, so S_3 is not a p-group.

If G is a group of order p^n, then it follows from the Lagrange's Theorem that order of every element of G is of the form p^m for some integer $m \geq 0$. Thus every group of order p^n is a p-group. Now we show, by the Cauchy's Theorem, that the converse also holds. Thus we have the following necessary and sufficient condition for a finite group to be a p-group.

Theorem: Let G be a finite nontrivial group. Then G is a p-group if and only if $|G| = p^n$ for some positive integer n.

Proof: First assume that G is a p-group. To prove $|G| = p^n$, it is sufficient to show that p is the only prime divisor of $|G|$. Let q be a prime divisor of $|G|$. Then G has an element a of order q, by the Cauchy's Theorem. Since G is a p-group, $q = p^r$ for some $r \geq 0$. It follows that $r = 1$ and $q = p$, since both p and q are primes.

Conversely, let $|G| = p^n$ where n is a positive integer. Since for every $a \in G$, $o(a) \mid p^n$ by the Lagrange's Theorem, so $o(a) = p^m$ for some $m \geq 0$. Therefore G is a p-group.

In the next theorem, we prove that the center of a p-group is nontrivial.

Theorem: Let G be a finite p-group such that $|G| > 1$. Then $|Z(G)| > 1$.

Proof: Since G is a finite p-group, so $|G| = p^n$ for some $n \geq 1$. Consider the class equation of G:

$$|G| = |Z(G)| + \sum_{a \notin Z(G)} [G : C(a)].$$

If $G = Z(G)$, then the result is trivial. Suppose $G \neq Z(G)$. Consider $a \in G \setminus Z(G)$. Then C(a) is a proper subgroup of G which implies, by the Lagrange's Theorem, that $|C(a)| = p^m$ for some $m < n$. Hence $p \mid [G : C(a)]$ for all $a \notin Z(G)$, and so p divides $\sum_{a \in Z(G)} [G : C(a)]$. Then it follows from the class equation that $p \mid |Z(G)|$. Hence $|Z(G)| > 1$. This result can be applied to get more precise inference on the groups of order p^n when $n = 2$ or 3.

Corollary: Let p be a prime number. Then every group of order p^2 is abelian.

Proof: Let G be a group of order p^2. Hence $|Z(G)| > 1$. Then, by Lagrange's theorem, $|Z(G)| = p$ or p^2. Now we show that $|Z(G)|$ can net be p. Otherwise, for $a \in G \setminus Z(G)$, Z(G) is a proper subset of C(a). Then $|C(a)| = p^2$ which implies that $G = C(a)$. This contradicts that $a \notin Z(G)$. Thus $|Z(G)| = p^2$, forcing $G = Z(G)$. Hence G is abelian.

Corollary: If G is a noncommutative group of order p^3 for some prime p, then $|Z(G)| = p$.

Proof: Since G is noncommutative, so $G \neq Z(G)$ which implies that $|Z(G)| \neq p^3$. Also, G being a nontrivial p-group, $|Z(G)| > 1$. Then the Lagrange's Theorem implies that $|Z(G)| = p$ or p^2. If $|Z(G)| = p^2$, then for every $a \in G \setminus Z(G)$, Z(G) is a proper subset of C(a) which implies that $|C(a)| = p^3$, that is $C(a) = G$ leading to the contradiction that $a \in Z(G)$. Thus it follows that $|Z(G)| = p$.

Another Version of Cauchy's Theorem

The Extended Complex Plane

Let $S = \{(x_1, x_2, x_3) \in \mathbb{R}^3 : x_2^1 + x_2^2 + (x_3 - 1/2)^2 = 1/4\}$. Thus S is the sphere in R^3 (called the Riemann sphere) with center at (0, 0, 1/2) and radius 1/2. The line segment joining (0, 0, 1), the north pole of S, toapoint(x, y, 0) is $\{(t_x, t_y, 1-t) : 0 \le t \le 1\}$, and this segment meets S when and only when,

$$t^2\left(x^2+y^2\right)+\left(\frac{1}{2}-t\right)^2 = \frac{1}{4}, \text{ or } t = \frac{1}{1+x^2+y^2}.$$

Therefore the intersection point is (x_1, x_2, x_3), where,

$$x_1 = \frac{x}{1+x^2+y^2}, x^2 = \frac{y}{1+x^2+y^2}, x^3 = \frac{x^2+y^2}{1+x^2+y^2}.$$

Since $1 - x_3 = 1/(1 + x^2 + y^2)$, it follows from equation above that,

$$x = \frac{x_1}{1-x_3}, y = \frac{x_2}{1-x_3},$$

Let h be the mapping that takes (x, y, 0) to the point (x_1, x_2, x_3) of S. Then h maps $\mathbb{R}^2 \times \{0\}$ which can be identified with $\mathbb{C}$, one-to-one onto $S\backslash\{(0, 0, 1)\}$. Also, by equation above,

$$h^{-1}\left(x_1, x_2, x_3\right) = \left(\frac{x_1}{1-x_3}, \frac{x_2}{1-x_3}, 0\right).$$

Consequently, h is a homeomorphism, that is, h and h^{-1} are continuous.

We can identify $\mathbb{C}$ and $S \backslash\{(0, 0, 1)\}$ formally as follows. Define $k : \mathbb{C} \to R^2 \times \{0\}$ by $k(x + iy) = (x, y.0)$. Then k is an isometry (a one-to-one, onto, distance-preserving map), hence $h \circ k$ is a homeomorphism of $\mathbb{C}$ onto $S \backslash\{(0, 0, 1)\}$. Next let ∞ denote a point not belonging to $\mathbb{C}$, and take $\hat{\mathbb{C}}$ to be $\mathbb{C} \cup \{\infty\}$. Define $g : \hat{\mathbb{C}} \to S$ by,

$$g(z) = \begin{cases} h(k(z)), & z \in \mathbb{C} \\ (0,0,1), & z = \infty. \end{cases}$$

Then g maps $\mathbb{C}$ one-to-one onto S. If ρ is the usual Euclidean metric of $\mathbb{R}^3$ and $\hat{d}$ is defined on $\hat{\mathbb{C}} \times \hat{\mathbb{C}}$ by,

$$\hat{d}(z,w) = p(g(z), g(w)),$$

then $\hat{d}$ is a metric on $\hat{\mathbb{C}}$. (The $\hat{d}$-distance between points of $\hat{\mathbb{C}}$ is the Euclidean distance between the corresponding points on the Riemann sphere.) The metric space $(\hat{\mathbb{C}}, \hat{d})$ is called the extended plane, and $\hat{d}$ is called the chordal metric on $\hat{\mathbb{C}}$. It is a consequence of the definition of $\hat{d}$ that $(\hat{\mathbb{C}}, \hat{d})$ and (S, ρ) are isometric spaces. The following formulas for $\hat{d}$ hold.

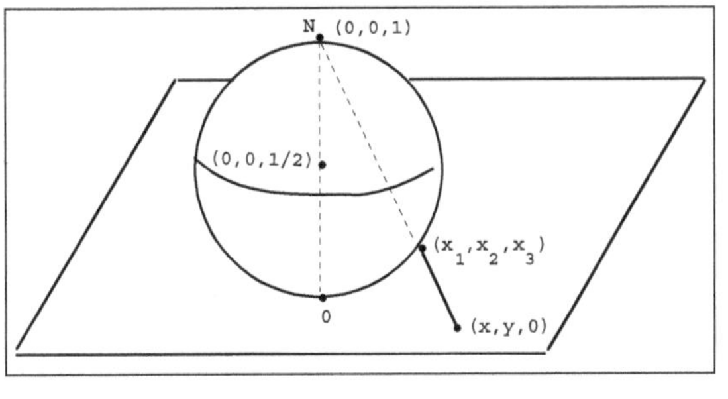

Lemma:

$$\hat{d}(z,w)=\begin{cases}\dfrac{|z-w|}{\left(1+|z|^2\right)^{1/2}\left(1+|w|^2\right)^{1/2}}, & z,w\in\mathbb{C}\\[2ex] \dfrac{1}{\left(1+|z|^2\right)^{1/2}} & z\in\mathbb{C}, w=\infty.\end{cases}$$

Proof: Suppose z = x + iy, w = u + iv. Then by equation,

$$x_1=\frac{x}{1+x^2+y^2}, x^2=\frac{y}{1+x^2+y^2}, x^3=\frac{x^2+y^2}{1+x^2+y^2}.$$

$$\left[\hat{d}(z,w)\right]^2=\left[\frac{x}{1+|z|^2}-\frac{u}{1+|w|^2}\right]^2+\left[\frac{y}{1+|z|^2}-\frac{v}{1+|w|^2}\right]^2+\left[\frac{|z|^2}{1+|z|^2}-\frac{|w|^2}{1+|w|^2}\right]^2$$

$$=\frac{x^2+y^2+|z|^4}{\left(1+|z|^2\right)^2}+\frac{u^2+v^2+|w|^4}{\left(1+|w|^2\right)^2}-2\frac{xu+yv+|z|^2|w|^2}{\left(1+|z|^2\right)^2\left(1+|w|^2\right)}$$

$$=\frac{|z|^2}{1+|z|^2}+\frac{|w|^2}{1+|w|^2}-\frac{|z|^2+|w|^2-|z-w|^2+2|z|^2|w|^2}{\left(1+|z|^2\right)\left(1+|w|^2\right)}$$

$$=\frac{|z-w|^2}{\left(1+|z|^2\right)\left(1+|w|^2\right)}.$$

as desired. Also,

$$\left[\hat{d}(z,\infty)\right]^2=\left[\rho\left(g(z),(0,0,1)\right)\right]^2$$

$$=\frac{x^2+y^2+1}{\left(1+x^2+y^2\right)^2}.$$

by equation,

$$x_1=\frac{x}{1+x^2+y^2}, x^2=\frac{y}{1+x^2+y^2}, x^3=\frac{x^2+y^2}{1+x^2+y^2}$$

$$=\frac{1}{1+|z|^2}.$$

Theorem:

- The metric space $(\hat{\mathbb{C}},\hat{d})$ is compact, and the identity function on $\mathbb{C}$ is a homeomorphism of $\mathbb{C}$ (with the usual metric) onto $(\mathbb{C}, \hat{d})$.

- The complex plane is a dense subspace of $\hat{\mathbb{C}}$. In fact, a sequence $\{z_n\}$ in $\mathbb{C}$ converges to ∞ iff $\{|z_n|\}$ converges to $+\infty$.
- The metric space $(\hat{\mathbb{C}}, \hat{d})$ is connected and complete.
- Let γ be a closed curve in $\mathbb{C}$, and define $n(\gamma,\infty) = 0$. Then the function $n(\gamma, \cdot)$ is continuous on $\hat{\mathbb{C}} \setminus \gamma^*$.
- The identity map on $\hat{\mathbb{C}}$ is a homeomorphism of $(\hat{\mathbb{C}}, \hat{d})$ with the one-point compactification $(\mathbb{C}_\infty, T)$ of $\mathbb{C}$. (Readers unfamiliar with the one-point compactification of a locally compact space may simply ignore this part of the theorem, as it will not be used later).

Proof: Since the Riemann sphere is compact, connected and complete, so is $(\hat{\mathbb{C}}, \hat{d})$. The formula for $\hat{d}$ in above lemma shows that the identity map on $\mathbb{C}$ is a homeomorphism of $\mathbb{C}$ into $\hat{\mathbb{C}}$ and that $z_n \to \infty$ iff $|z_n| \to +\infty$. We are now going to make precise, in two equivaxalent ways, the notion that an open set has no holes.

Theorem: Let Ω be open in $\mathbb{C}$. Then $\hat{\mathbb{C}} \setminus \Omega$ is connected iff each closed curve (and each cycle) γ in Ω is Ω-homologous to 0, that is, $n(\gamma, z) = 0$ for all $z \notin \Omega$.

Proof: Suppose first that $\hat{\mathbb{C}} \setminus \Omega$ is connected, and let γ be a closed curve in Ω. Since $z \to n(\gamma, z)$ is a continuous integer-valued function on $C \setminus \gamma*$, it must be constant on the connected set $\hat{\mathbb{C}} \setminus \Omega$. But $n(\gamma,\infty) = 0$, hence $n(\gamma, z) = 0$ for all $z \in \hat{\mathbb{C}} \setminus \Omega$. The statement for cycles now follows from the result for closed curves.

The converse is considerably more difficult, and is a consequence of what we will call the hexagon lemma. This lemma has several applications in addition to its use in the proof of the converse.

Completion of the Proof of Theorem

If $\hat{\mathbb{C}} \setminus \Omega$ is not connected, we must exhibit a cycle in Ω that is not Ω-homologous to 0. Now since $\hat{\mathbb{C}}$ is closed and not connected, it can be expressed as the union of two nonempty disjoint closed sets K and L. One of these two sets must contain ∞; assume that $\infty \in L$.

Then K must be a compact subset of the complex plane $\mathbb{C}$, and K is contained in the plane open set $\Omega_1 = C \setminus L$. apply the hexagon lemma to Ω_1 and K to obtain a cycle σ in $\Omega_1 \setminus K = \mathbb{C} \setminus (K \cup L) = \Omega$ such that $n(\sigma, z) = 1$ for each $z \in K$ (and $n(\sigma, z) = 0$ for $z \notin \Omega_1$). Pick any point z in the nonempty set $K \subseteq \mathbb{C} \setminus \Omega$. Then $z \notin \Omega$ and $n(\sigma, z)=1 \neq 0$.

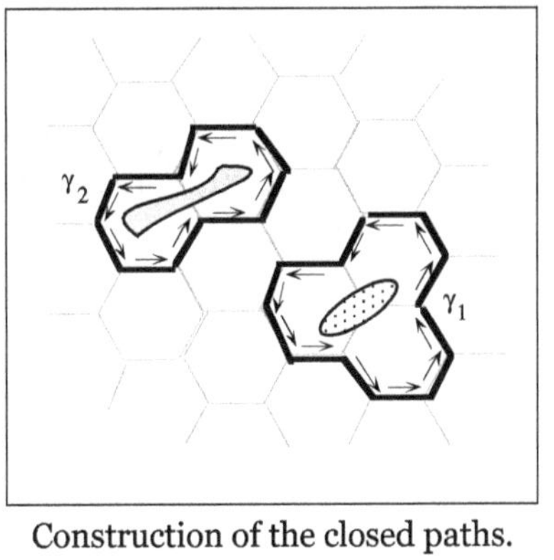

Construction of the closed paths.

Second Cauchy Theorem

- Let Ω be an open subset of $\mathbb{C}$. The following are equivalent.
- $\hat{\mathbb{C}} \setminus \Omega$ is connected.
- $n(\gamma, z) = 0$ for each closed path (or cycle) γ in Ω and each point $z \in \mathbb{C} \setminus \Omega$.
- $\int_\gamma f(z)dz = 0$ for every closed path (or cycle) γ in Ω and every analytic function f on Ω.
- Every analytic function on Ω has a primitive on Ω.
- Every zero-free analytic function on Ω has an analytic logarithm.
- Every zero-free analytic function on Ω has an analytic n-th root for n =1, 2,...

It is true that in complex analysis, the implications (1) ⇒ (2) ⇒ (3) are used almost exclusively. The rather tedious hexagon lemma was required to establish the reverse implication (2) ⇒ (1). Thus one might wonder why we have gone to the trouble of obtaining the hexagon lemma at all. One answer is that it has other applications, including the following global integral representation formula.

Theorem: Let K be a compact subset of the open set Ω. Then there is a cycle γ in $\Omega \setminus K$ such that γ is a formal sum of closed polygonal paths, and for every analytic function f on Ω,

$$f(z) = \frac{1}{2\pi i}\int_\gamma \frac{f(w)}{w-z}dw = 0 \quad \text{for all } z \in K.$$

The Hexagon Lemma

Let Ω be an open subset of $\mathbb{C}$, and let K be a nonempty compact subset of Ω. Then there are closed polygonal paths $\gamma_1, \gamma_2, \ldots, \gamma_m$ in $\Omega \setminus K$ such that,

$$\sum_{j=1}^{m} n(\gamma_j, z) = \begin{cases} 1 & \text{if } z \in K \\ 0 & \text{if } z \notin \Omega. \end{cases}$$

The lemma may be expressed by saying there is a (polygonal) cycle in $\Omega \setminus K$ which winds around each point of K exactly once, but does not wind around any point of $\mathbb{C} \setminus \Omega$.

Proof: For each positive integer n, let P_n be the hexagonal partition of $\mathbb{C}$ determined by the hexagon with base [0, 1/n. Since K is a compact subset of the open set Ω, we have dist$(K, \mathbb{C} \setminus \Omega) > 0$, and therefore we can choose n large enough so that if $H \in P_n$ and $H \cap K \neq \emptyset$, then $H \subseteq \Omega$. Define $\mathbb{K} = \{H \in P_n : H \cap K \neq \emptyset\}$. Since K is nonempty and bounded, $\mathbb{K}$ is a nonempty finite collection and,

$$K \subseteq \cup\{H : H \in \mathcal{K}\} \subseteq \Omega.$$

Now assign a positive (that is, counterclockwise) orientation to the sides of each hexagon. Let S denote the collection of all oriented sides of hexagons in $\mathcal{K}$ that are sides of exactly one member of $\mathcal{K}$. Observe that given an oriented side $\vec{ab} \in S$, there are unique oriented sides $\overrightarrow{ca}$ and $\overrightarrow{bd}$ in S.

(This uniqueness property is the motivation for tiling with hexagons instead of squares. If we used squares instead, we have $\vec{ab}$, $\vec{bc}$ and $\vec{bd} \in S$, so $\vec{ab}$, does not have a unique successor, thus complicating the argument that follows).

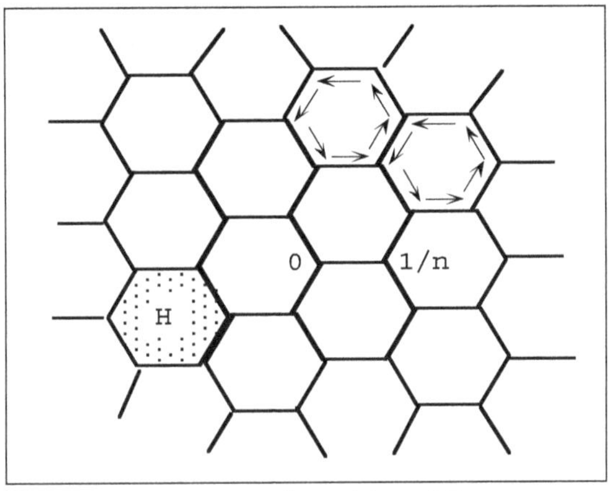

A Hexagonal Partition of ℂ.

By the above observations, and the fact that S is a finite collection, it follows that given $a_1 \vec{a}_2 \in S$ there is a uniquely defined closed polygonal path $\gamma_1 = [a_1, a_2, ..., a_k, a_1]$ with all sides in S. If S_1 consists of the edges of γ_1 and $S \backslash S_1 \neq \emptyset$, repeat the above construction with S replaced by $S\backslash S_1$. Continuing in this manner, we obtain pairwise dis-joint collections $S_1, S_2, ..., S_n$ such that $S = \cup_{j=1}^{m} S_j$, and corresponding closed polygonal paths $\gamma_1, \gamma_2, ..., \gamma_m$.

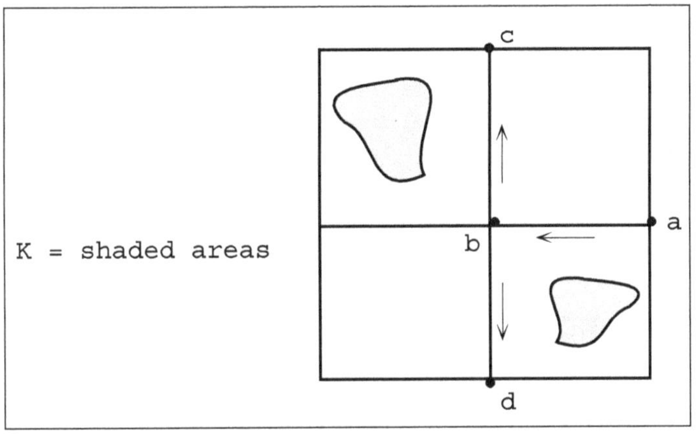

Non-uniqueness when squares are used.

Suppose now that the hexagons in $\mathbb{K}$ are $H_1, H_2, ..., H_p$, and let σ_j denote the boundary of H_j, oriented positively. If z belongs to the interior of some H_r, then $n(\sigma_r, z) = 1$ and $n(\sigma_j, z) = 0, j \neq r$. Consequently, $n(\sigma_1 + \sigma_2 + \cdots + \sigma_p, z) = 1$. But by construction, $n(\gamma_1 + \cdots + \gamma_m, z) = n(\sigma_1 + \cdots + \sigma_p, z)$. (The key point is that if both hexagons containing a particular side [a, b] belong to $\mathbb{K}$, then $\vec{ab} \notin S$, and $\vec{ba} \notin S$, Thus [a, b] will not contribute to either $n(\gamma_1 + \cdots + \gamma_m, z)$ or to $n(\sigma_1 + \cdots + \sigma_p, z)$. If only one hexagon containing [a, b] belongs to $\mathbb{K}$, then $\vec{ab}$ (or $\vec{ba}$) appears in both cycles). Therefore $n(\gamma_1 + \cdots + \gamma_m, z) = 1$. Similarly, if $z \notin \Omega$, then $n(\gamma_1 + \cdots + \gamma_m, z) = n(\sigma_1 + \cdots + \sigma_p, z) = 0$.

Finally, assume $z \in K$ and z belongs to a side s of some H_r. Then s cannot be in S, so $z \notin (\gamma_1 + \cdots + \gamma_m)^*$. Let $\{w_k\}$ be a sequence of interior points of H_r with w_k converging to z. We have shown that $n(\gamma_1 + \cdots + \gamma_m, w_k) = 1$ for all k, so, $n(\gamma_1 + \cdots + \gamma_m, z) = 1$.

Sylow's Theorem

The Sylow theorems are important tools for analysis of special subgroups of a finite group G, known as Sylow subgroups. They are especially useful in the classification of finite simple groups. The first Sylow theorem guarantees the existence of a Sylow subgroup of G for any prime p dividing the order of G. A Sylow subgroup is a subgroup whose order is a power of p and whose index is relatively prime to p. The second Sylow theorem states that all the Sylow subgroups of a given order are conjugate, and the third Sylow theorem gives information about the number of Sylow subgroups.

The proofs of the theorems use nontrivial facts about group actions, in particular the action of G on the coset space G/H, where H has prime power order.

The Theorems

Theorem: Let G be a finite group. Let p be a prime dividing $|G|$. Write $|G|=p^k m$ with $k \geq 1$ and $p \nmid m$.

- First Sylow Theorem: There is a subgroup $H \subseteq G$ of order p^k. H is called a Sylow p-subgroup.
- Second Sylow Theorem: Any two Sylow p-subgroups are conjugate: if H and K are Sylow p-subgroups, there is an element $g \in G$ such that $g^{-1}Hg = K$.
- Third Sylow Theorem: Let n_p be the number of Sylow p-subgroups. Then,
    - $n_p|m$.
    - $n_p \equiv 1 (\text{mod } p)$.
    - $n_p = |G|/|N_G(H)|$, where H is any Sylow p-subgroup and $N_G(H)$ denotes the normalizer of H, the largest subgroup of G in which H is normal.

Examples and Applications

Example: Identify the Sylow subgroups of S_4.

$$|S_4| = 24 = 2^3 \cdot 3.$$

The Sylow 2-subgroups have order 8. There is a natural example of an order-8 subgroup in S_4, namely the dihedral group D_4. Labeling the vertices of a square 1, 2, 3, 4, the dihedral group is the set of symmetries of the square. It is generated by a 90-degree rotation, which is a 4-cycle, and a flip, which is a double transposition. There are four flips and four rotations (including the identity).

The dihedral group is not normal inside S_4; labeling the vertices differently leads to different copies. For instance, one copy is generated by (13)(24) and (1234), but another copy is generated by (12)(34) and (1324) and a third copy is generated by (14)(23) and (1243). These are all conjugate, as predicted by the second Sylow theorem; and $n_2 = 3$ meets the criteria of the third Sylow theorem: $n_2 \equiv 1$ (mod2) and $n_2 | 3$.

The Sylow 3-subgroups are generated by the 3-cycles, like (123). There are 8 such 3-cycles, hence four subgroups (each subgroup contains two of them). So $n_3 = 4$. This agrees with the third Sylow theorem, since $n_3|8$ and $n_3 \equiv 1 \pmod 3$. As for $NS_4(H)$, the third Sylow theorem predicts that this is a subgroup of order $|S_4|/n_3 = 24/4 = 6$. Let $H = \langle(123)\rangle$ be one of the Sylow 3-subgroups, then $N_G(H)$ is a copy of S_3 inside S_4, consisting of all the permutations that fix 4. Any conjugate of a Sylow p-subgroup is a Sylow p-subgroup. So if $n_p = 1$, so that the Sylow p-subgroup is unique, then it must be normal. This is a common source of applications.

Example: Show that there is no simple group of order 30. Suppose G is a simple group of order 30. The third Sylow theorem for p = 3 says that $n_3|10$ and $n_3 \equiv 1 \pmod 3$. So $n_3 = 1_n$ or 10. If $n_3 = 1$ then the Sylow 3-subgroup is normal, which is impossible since G is simple. So $n_3 = 10$ and there are 10 Sylow 3-subgroups. They are distinct, so they intersect only in the identity element (the intersection is a subgroup, and by Lagrange's theorem its order is 1 or 3, so it must be 1). Any Sylow 3-subgroup has order 3, and consists of the identity and two elements of order 3. So there are 20 elements of order 3 in G.

Now $n_5|6$ and $n_5 \equiv 1 \pmod 5$, so $n_5 = 1$ or 6. As before, n_5 must be 6, so there are 6 Sylow 5-subgroups with 5 elements each, consisting of the identity and four elements of order 5. Any pair of subgroups intersects only in the identity, so there are a total of 24 elements of order 5 in G. But this count, gives at least 20 + 24 = 44 elements in a group of order 30, which is impossible. So there is no simple group of order 30.

First Theorem

The French mathematician M. L. Sylow proved three results on the structure of finite groups, which are applied to characterize simple groups. It is interesting to note that these results were proved first by M. L. Sylow for permutation groups. It is the Cayley's theorem which influenced George Frobenious to prove the Sylow Theorems in the general setting.

Lemma: Let G be a group of order p^n, p be a prime and S be a finite G-set. If $S_0 = \{a \in S | ga = a \text{ for all } g \in G\}$, then,

$$|S| \equiv |S_0| \pmod p.$$

Proof: Let $a \in S$. Then $a \in S_0$ if and only if the orbit $[a] = \{a\}$, equivalently $|[a]| = 1$. Hence S can be written as a disjoint union $S = S_0 \cup [a_1] \cup [a_2] \cup \cdots \cup [a_k]$, where $[a_1], [a_2], \cdots, [a_k]$ are distinct orbits of G on S and disjoint with S_0. Then $|S| = |S_0| + |[a_1]| + \cdots + |[a_k]|$.

Since $a_i \notin S_0$, so there is $g \in G$ such that $g_{a_i} \neq a_i$ which implies that $g \notin G_{a_i}$ and so $|G_{a_i}| < |G|$. Thus $|[a_i]| > \dfrac{|G|}{|G_{a_i}|} > 1$. Then $|[a_i]| > 1$ and $|[a_i]| = [G:G_{a_i}]| \; |G| = p^n$ imply that $p \mid |[a_i]|$ for all $1 \leq i \leq k$.

Hence $p \left| \sum_{i=1}^{k} |[a_i]| \right.$ and it follows that $|S| \equiv |S_0| \pmod p$.

Recall that if every element of a group G has order p^n; $n \geq 0$ for some fixed prime p, then G is called a p-group. With the help of Cauchy's Theorem we have proved that a finite group G is a p-group if and only if $|G| = p^n$ for some $n \geq 0$.

We call a subgroup H of a group G a p-subgroup if H is a p-group. In particular, for every prime p, {e} is a p-subgroup of every group G, since $|\{e\}| = 1 = p^0$.

Lemma: If H is a p-subgroup of a finite group G, then $[N_G(H) : H] \equiv [G : H] (\text{mod } p)$.

Proof: Let S be the set of all left cosets of H in G. Then $H \times S \mapsto S$ given by $(g, aH) \mapsto (ga)H$ is an action of H on S. Since H is a p-group, $|S| \equiv |S_0| (\text{mod } p)$.

Here $|S| = [G : H]$. Now for $a \in G$,

$$\begin{aligned} aH \in S_0 &\Leftrightarrow gaH = aH \text{ for all } g \in G, \\ &\Leftrightarrow a^{-1} ga \in H \text{ for all } g \in G, \\ &\Leftrightarrow a^{-1} Ha \subseteq H, \\ &\Leftrightarrow a \in N_G(H). \end{aligned}$$

shows that S_0 is the set of all left cosets aH such that $a \in N_G(H)$, that is $|S_0| = [N_G(H) : H]$. Hence $[G : H] \equiv [N_G(H) : H] (\text{mod } p)$.

Corollary: Let p be a prime and H be a p-subgroup of a finite group G. If $p \mid [G : H]$, then $N_G(H) \neq H$.

Proof: Since G is a finite group and H is a p-subgroup of G, so,

$$[G:H] \equiv [N_G(H):H] (\text{mod } p).$$

Then $p \mid [G : H]$ implies that $p \mid [N_G(H) : H]$. Since $[N_G(H) : H] \neq 0$ it follows that $[N_G(H) : H] \geq p$ and the result follows.

Theorem: (Sylow's First Theorem) - Let G be a finite group of order $p^n m$, where p is a prime, $n > 0$ and $\gcd(p, m) = 1$. Then G has a subgroup of order p^i for every $1 \leq i \leq n$. Moreover, for every subgroup of order p^i, $i < n$, there is a subgroup K of order p^{i+1} such that H is normal in K.

Proof: We prove that G has a subgroup H of order p^i for all $1 \leq i \leq n$ by induction on i. Since $n > 0$, $p \mid |G|$ and it follows from the Cauchy's Theorem that G has a subgroup H of order p. Assume that H is a subgroup of order p^i, where $1 \leq i < n$. Since H is a subgroup of order p^i, so $[G : H] \equiv [N(H) : H] (\text{mod } p)$. Then $p \mid [G : H]$ implies that $p \mid [N(H) : H] = |N(H)/H|$, and it follows from the Cauchy's Theorem that the quotient group $N(H)/H$ has a subgroup K/H of order p. Thus G has a subgroup K such that $|K| = |K/H| \cdot |H| = p^{i+1}$. Hence, by the principle of mathematical induction, G has a subgroup of order p^i for every $1 \leq i \leq n$. Finally, $K \subseteq N(H)$ shows that H is normal in K. As an immediate consequence we have following extension of Cauchy's Theorem:

Corollary: Let G be a finite group and p be a prime. If $p^n \mid |G|$, then G has a subgroup H of order p^n.

The second part of the Sylow's first theorem motivates us to introduce the notion of maximal p-subgroups.

Definition: Let p be a prime, then a subgroup P of G is called a Sylow p-subgroup if it is a maximal p-subgroup of G, that is, if H is a p-subgroup of G and $P \subseteq H \subseteq G$ then $P = H$.

If G is a finite group then G can have only finitely many subgroups and a maximal p-subgroup. If G is infinite then, also G has a maximal p-subgroup, by the Zorn's Lemma. Thus, for every prime p every group G has a Sylow p-subgroup, though they may be trivial. We have following equivalent characterization of Sylow p-subgroups.

Theorem: Let G be a finite group of order $p^n m$, $n > 0$ and $\gcd(p, m) = 1$. Then a subgroup H is a Sylow p-subgroup of G if and only if $|H| = p^n$.

Proof: First assume that H is a subgroup of G such that $|H| = p^n$. Let K be a p-subgroup of G such that $H \subseteq K$. Since K is a p-subgroup, so $|K| = p^r$ for some $r \geq 0$. Now by the Lagrange's Theorem, $|K| \mid |G|$ i.e. $p^r \neq p^n m$ and $\gcd(p, m) = 1$ together implies that $r \leq n$. Also $H \subseteq K$ implies that $n \leq r$. Thus $r = n$ and $|H| = |K| = p^n$. Hence $H = K$ and H is a maximal p-subgroup.

Conversely, suppose that H is a Sylow p-subgroup. Then $|H| = p^r$ for some $r \geq 0$. Now, by the Lagrange's Theorem, $|H| \mid |G|$, that is $p^r| p^n m$ and $\gcd(n, m) = 1$ implies that $r \leq n$. If $r < n$ then, the Sylow's first theorem implies that H is contained in a subgroup K of order p^{r+1} which contradicts that H is a maximal p-subgroup of G. Thus $r = n$ and $|H| = p^n$. Following result has an important role to characterize the nilpotent groups:

Theorem: Let G be a finite group. If every Sylow subgroup of G is normal, then G is a direct product of its Sylow subgroups.

Proof: Let $|G| = p_1^{n_1} p_2^{n_2} \cdots p_k^{n_k}$, where p_i are distinct primes. Since each Sylow subgroups of G is normal, so G has unique Sylow p_i-subgroup for every p_i, say P_i. Let $a_i \in P_i$ and $a_j \in P_j$, where $i \neq j$. Since both P_i and P_j are normal, so $a_i a_j a_i^{-1} a_j^{-1} \in P_i \cap P_j$. Now $\gcd(|P_i|, |P_j|) = 1$ implies, by the Lagrange's Theorem, that $|P_i \cap P_j| = 1$ and so $P_i \cap P_j = \{e\}$. Thus $a_i a_j = a_j a_i$. Now let $a \in P_r \cap (P_1 \cdots P_{r-1} P_{r+1} \cdots P_k)$. Then,

$$a = a_1 \cdots a_{r-1} a_{r+1} \cdots a_k,$$

where $a_i \in P_i$. Now $o(a_i) \mid |P_i|$, that is $o(a_i) \mid p_i^{n_i}$. Thus $o(a) \mid p_r^{n_r}$ and $a_i a_j = a_j a_i$ for all $i \neq j$ implies that $o(a) \mid p_1^{n_1} \cdots p_{r-1}^{n_{r-1}} p_{r+1}^{n_{r+1}} \cdots p_k^{n_k}$. Since $\gcd\left(p_r^{n_r}, p_1^{n_1} \cdots p_{r-1}^{n_{r-1}} p_{r+1}^{n_{r+1}} \cdots p_k^{n_k}\right) = 1$, so $o(a) = 1$ and $a = e$. Thus $P_r \cap (P_1 \cdots P_{r-1} P_{r+1} \cdots P_k) = \{e\}$ which implies that $|P_1 P_2 \cdots P_k| = |P_1||P_2| \cdots |Pk| = p_1^{n_1} p_2^{n_2} \cdots p_k^{n_k} = |G|$. Hence $G = P_1 P_2 \cdots P_k$ and G is an internal direct product of the Sylow subgroups $P_1, P_2, \cdots, P_k$.

Second Theorem

Let P be a Sylow p-subgroup of the finite group G.

Let Q be any p-subgroup of G.

Then Q is a subset of a conjugate of P.

Proof: Let P be a Sylow p-subgroup of G.

Let $\mathbb{S}$ be the set of all distinct G-conjugates of P:

$$\mathbb{S} = \left\{ gPg^{-1} : g \in G \right\}.$$

Let h $*$ S be the conjugacy action:

$$\forall h \in P, S \in \mathbb{S} : h * S = hSh^{-1}.$$

From Conjugacy Action on Subgroups is Group Action, this is a group action for $S \le G$.

To show it is closed for $S \in \mathbb{S}$:

$$\begin{aligned} & S \in \mathbb{S} \\ \leadsto \quad & \exists g \in G : S = gPg^{-1} \\ \leadsto \quad & h * S = h\left(gPg^{-1}\right)h^{-1} \\ & \quad = (hg)P(hg)^{-1} \\ \leadsto \quad & h * S \in \mathbb{S}. \end{aligned}$$

So, consider the orbits and stabilizers of $\mathbb{S}$ under this group action.

Since $\forall S \in \mathbb{S}$: Stab (S) $\le$ P, we have that:

$$|\text{Stab}(S)| \setminus |P|.$$

Therefore, by the Orbit-Stabilizer Theorem, these orbit lengths are all congruent to either 0 or 1 modulo p, since P is a Sylow p-subgroup of G.

This will imply, as we shall mark later on:

$$|\mathbb{S}| \equiv 1 \pmod{p}.$$

Now, h $*$ P = hPh^{-1} = P, so:

Orb (P) = {P}.

We now show that P is the only element of $\mathbb{S}$ such that |Orb(S)|=1.

If gPg^{-1} has one element in its orbit, then:

$$\forall x \in P : x\left(gPg^{-1}\right)x^{-1} = gPg^{-1}.$$

Thus, $\forall x \in P$ we have that:

$$g^{-1}xg \in N_G(P).$$

From Order of Conjugate Element equals Order of Element, we have that:

$$\left|g^{-1}xg\right| = |x|.$$

Thus, $P_1 = g^{-1} Pg$ is a p-subgroup of $N_G(P)$.

As P and P_1 have the same number of elements, P_1 is a Sylow p-subgroup of $N_G(P)$.

Hence: $P_1 = P$ by Normalizer of Sylow p-Subgroup, so $gPg^{-1} = P$.

Thus, P is the only element of $\mathbb{S}$ whose orbit has length 1.

From Stabilizer of Coset Action on Set of Subgroups, $P = N_G(P)$.

Thus, for any $g \notin P$, $|\mathrm{Orb}(gPg^{-1})|$ under conjugation by elements of P has orbit greater than 1.

Hence: $|\mathbb{S}| \equiv 1 \pmod p$ as promised.

Next we consider orbits of $\mathbb{S}$ under conjugation by elements of Q.

Since every orbit has length a power of p, the above conclusion shows there is at least one orbit of length 1.

So there is an element g such that:

$$\forall x \in Q : x\left(gPg^{-1}\right)x^{-1} = gPg^{-1}.$$

As previously:

$$g^{-1}Qg \subseteq N_G(P).$$

So by Normalizer of Sylow p-Subgroup:

$$g^{-1}Qg \subseteq P.$$

Thus, $Q \subseteq gPg^{-1}$ as required.

Third Theorem

In the Sylow's third theorem, the number of Sylow p-subgroups of a group G has been characterized. Let S be the set of all Sylow p-subgroups. Since every conjugate of a Sylow p-subgroup is also a Sylow p-subgroup, so for every subgroup H of G,

$$H \times S \to S, \text{ given by}$$
$$(a, P) \mapsto aPa^{-1}.$$

is an action of H on S. Also any two Sylow p-subgroup are conjugate on S, which implies that if G acts on S by conjugation then there is single orbit of G on S.

Theorem: (Sylow's Third Theorem) - Let G be a finite group of order $p^n m$, where p is a prime, $n > 0$ and $\gcd(p, m) = 1$. Then the number n_p of Sylow p-subgroup is of the form $1 + kp$ for some $k \geq 0$ and is such that $n_p \mid p^n m$.

Proof: Let S be the set of all Sylow p-subgroups of G and P be a Sylow p-subgroup. Consider action of P on S by conjugation, that is,

$$(p, H) \mapsto pHp^{-1}.$$

Since P is a p-group, so $|S| \equiv |S_0| \pmod p$, where $S_0 = \{H \in S \mid pHp^{-1} = H \text{ for all } p \in P\} = \{H \in S \mid P \subseteq$

N(H)}. Then $P \in S_0$ and $S_0 \neq \varphi$. Let $Q \in S_0$. Then $P \subseteq N(Q)$. Since both P, Q are Sylow p-subgroups of G, so they are Sylow p-subgroups of N(Q) and are conjugate in N(Q). Let $a \in N(Q)$ such that $P = aQa^{-1}$. But then P = Q, since $aQa^{-1} = Q$. Hence $S_0 = \{P\}$ and it follows that $|S| \equiv 1 (\text{mod } p)$, that is $|S| = 1 + kp$ for some $k \geq 0$.

To prove the second part, consider action of G on S by conjugation. Then there is only one orbit of G on S, and so S = [P] for every Sylow p-subgroup P of G. Thus,

$$n_p = |S| = |[P]| = [G : G_p].$$

and hence $n_p \mid |G|$.

Now we give some examples to show the applications of Sylow's Theorems to study structure of finite groups.

Example: Let G be a group of order $45 = 3^2 \cdot 5$. Denote the number of Sylow 3-subgroups of G by n_3. Then $n_3 = 3k + 1$ for some $k \geq 0$ and $n_3 \mid 45$. Hence $n_3 = 1$ and G has unique Sylow 3-subgroup H. Thus H is a normal subgroup of G of order $3^2 = 9$.

Example: We show that every group of order of 99 is abelian.

Let G be a group of order $99 = 11 \cdot 3^2$. Let n_3 be the number of Sylow 3-subgroups of G. Then $n_3 = 3k + 1$ for some integer $k \geq 0$ and $n_3 | 99$. It follows that G has unique Sylow 3-subgroup, say H which is normal in G and $|H| = 9$. Similarly G has unique Sylow 11-subgroup, say K which is normal in G and $|K| = 11$. Then $H \cap K = \{e\}$ and $|HK| = |H||K| = 99$ implies that G = HK. Thus G is an internal direct product of H and K. Hence $G \cong H \times K$. Since $|H| = 3^2$, so H is abelian and $|K| = 11$ implies that K is abelian. Therefore G is abelian. Also note that $G \cong \mathbb{Z}_{99}$ or $Z_3 \oplus Z_{33}$.

Example: We show that every group of order 15 is cyclic.

Let G be a group of order $15 = 3 \times 5$. Denote the number of Sylow 3-subgroups of G by n_3. Then $n_3 = 3k + 1$ for some $k \geq 0$ and $n_3 \mid 15$. Hence $n_3 = 1$ and G has unique Sylow 3-subgroup, say H. Then H is normal in G. Similarly G has unique Sylow 5-subgroup K and so K is normal in G. Since gcd(|H|, |K|) = 1, so $|H \cap K| = 1$ and we have,

$$|HK| = \frac{|H||K|}{|H \cap K|} = 15.$$

Thus, G = HK and G becomes an internal direct product of H and K. Now $|H| = 3$, a prime implies that $H \cong Z_3$. Similarly $K \cong \mathbb{Z}_5$. Therefore,

$$G \cong H \times K \cong \mathbb{Z}_3 \times \mathbb{Z}_5 \cong \mathbb{Z}_{15}.$$

Example: We show that every group of order 255 is cyclic.

Let G be a group of order $255 = 17 \cdot 5 \cdot 3$. Let n_{17} be the number of Sylow 17-subgroups. Then $n_{17} = 17k + 1$ for some integer $k \geq 0$ and $n_{17}|255$ implies that $n_{17} = 1$. Thus, G has unique Sylow 17-subgroup, say H which is normal in G and $|H| = 17$. Then the normalizer of H is N(H) = G. Since $\frac{N(H)}{C(H)}$

is isomorphic to a subgroup of Aut(H), so $\left|\frac{N(H)}{C(H)}\right|$ divides $|Aut(H)| = |Aut(Z_{17})|$. Now $Aut(\mathbb{Z}_{17}) \cong U_{17}$ implies that $\left|\frac{G}{C(H)}\right|$ divides 16. Also $\left|\frac{G}{C(H)}\right| \Big| |G| = 255$. Then gcd(16, 255) = 1 implies that $\left|\frac{G}{C(H)}\right| = 1$ and $G = C(H)$. Thus every element of H commutes with every element of G, whence $H \subseteq Z(G)$. Then $17 \mid |Z(G)|$, which also divides 255. Hence $|Z(G)| = 17, 51, 85$ or 255 and so $\frac{G}{Z(H)} = 15, 5, 3$ or 1. But every group of order 15, 5, 3 or 1 is cyclic; and it follows that G is abelian. Hence, by the fundamental theorem of finite abelian groups, $G \cong Z_{17} \oplus \mathbb{Z}_5 \oplus Z_3 \cong Z_{255}$. Now we characterize all groups of order 2p, where p is a prime.

Theorem: Let p be an odd prime. If G is a group of order 2p, then $G \cong Z_{2p}$ or $G \cong D_p$.

Proof: Let G be a group of order 2p. Then Cauchy's Theorem implies that G has an element a of order p and an element b of order 2. Denote H = < a >. Then [G : H] = 2 implies that H is normal in G. So $bab = bab^{-1} \in H$ which implies that $bab = a^i$ for some $0 \le i < p$. Then $a^{i^2} = (a^i)^i = (bab)^i = (bab^{-1})^i = ba^ib^{-1} = ba^ib$. Also $bab = a^i$ implies that $a = b^{-1}a^ib^{-1} = ba^ib$. Thus $a^{i^2} = a$, that is, $a^{i^2-1} = e$, and so $p \mid i^2 - 1 = (i - 1)(i + 1)$, since $\circ(a) = p$. Since p is a prime, so $p \mid i - 1$ or $p \mid i + 1$.

If $p \mid i - 1$, then $i < p$ implies that $i - 1 = 0$ that is $i = 1$. Then $bab = a$; and so $ba = ab^{-1} = ab$, forcing $\circ(ab) = lcm(\circ(a), \circ(b)) = lcm(2, p) = 2p$, since p is an odd prime. Hence $G \cong Z_2p$.

If $p \mid i + 1$, then $i = pk - 1$ for some integer k, and we have $bab = a^i = a^{-1} = a^{p-1}$, that is, $ba = a^{p-1}b^{-1} = a\ p^{-1}b$. Also we have H = < a > is normal in G such that $\left|\frac{G}{H}\right| = 2$. Since $\circ(b) = 2 \nmid p = |H|$ so $b \notin H$ and $bH \neq H$. Thus G/H = < H, bH >. Therefore every element of G can be written as $b^m a^n$ and so G = < a, b >. Since $\circ(a) = p$, $\circ(b) = 2$ and $ba = a^{p-1}b$, it follows that $G \cong D_p$.

Applications of Sylow's Theorem

The notion of simple group was introduced by Galois in his work on the insolvability of quintics. The simplicity of A_5 plays a crucial role in his proof that there are quintics which cannot be solved by radicals. But simple groups are known as the El Dorado of finite group theory because of their role to study structure of finite groups. Their role in finite group theory is somewhat analogous to that of primes in number theory, that is, they are the basic building blocks for all groups. Theory of simple groups is vast and difficult. Here we consider only simple groups of small order.

The Cauchy's Theorem shows that no abelian group of Composite order is simple. As a consequence of Cauchy's theorem we have no simple groups of order p^n, p is a prime and n > 1. Sylow theorems also help us to find possible orders of simple groups. Here we show that 60 is the smallest composite order of a noncommutative simple group and any noncommutative simple group of order 60 is isomorphic to A_5.

A group $G \neq \{e\}$ is called simple if its only normal subgroups are the identity subgroup and the group itself.

Example: We show that no group of order 10 is simple. Let G be a group of order $10 = 5 \times 2$. Denote the number of Sylow 5-subgroups of G by n_5. Then $n_5 = 5k + 1$ for some integer $k \geq 0$ and $n_5|10$. It follows that $n_5 = 1$, that is, G has unique sylow 5-subgroup H of order 5. Hence H is normal in G, and so G is not simple.

Theorem: No group of order pq, where p, q are prime integers, is simple.

Proof: Let G be a group of order pq. If p = q, then $|G| = p^2$ implies that G is abelian. Also, by Cauchy's Theorem, G has an element a and hence a subgroup $H = \langle a \rangle$ of order p. Since G is abelian, H is normal in G. Thus G is not simple.

Let p > q. Let n_p be the number of Sylolw p-subgroups of G. Then $n_p = kp + 1$ for some integer $k \geq 0$ and $n_p|pq$. Now gcd (n_p, p) = gcd $(kp + 1, p) = 1$ implies that $n_p|q$. Since p > q, so $n_p = kp + 1 \leq q$ holds only if k = 0, that is, $n_p = 1$. Thus G has unique Sylow p-subgroup H of order p, which is normal in G. Hence G is not simple.

Example: We show that no group of order 30 is simple. Let G be a group of order $30 = 5 \times 3 \times 2$. Let n_5 denote the number of Sylow 5-subgroups of G and n_3 denote the number of Sylow 3-subgroups of G. If $n_5 = 1$ or $n_3 = 1$ then G has a normal subgroup of order 5 or 3 and G is not simple. If possible, $n_5 \neq 1$ and $n_3 \neq 1$. Now $n_5 = 5k + 1$ for some integer $k \geq 0$ and $n_5|30$ implies that $n_5 = 6$(since $n_5 \neq 1$). Similarly $n_3 = 10$. Let $P_1, P_2, \cdots P_6$ be the six Sylow 5-subgroups. Then $|P_i| = 5$, a prime implies that $P_i \cap P_j = \{e\}$ for all $i \neq j$ and their union contains $6 \times 4 = 24$ elements of order 5. Similarly the 10 Sylow 3-subgroups contain $10 \times 2 = 20$ elements of order 3. Thus $|G| \geq 24 + 20 = 44$, a contradiction. Thus $n_5 = 1$ or $n_3 = 1$, and G is not simple. In fact, every group of order 30 has normal subgroups of order 5 and 3.

Theorem: Let p be a prime integer and n > 1 be an integer. Then no group of order p^n is simple.

Proof: Let G be a group of order p^n. Then $|Z(G)| > 1$. If G = Z(G) then G is abelian and so G cannot be simple, since n > 1. Let $G \neq Z(G)$. Then Z(G) becomes a nontrivial proper normal subgroup of G. Hence G is not simple.

Example: We show that no group of order 40 is simple. Let G be a group of order $40 = 5 \times 2^3$. Denote the number of Sylow 5-subgroups of G by n_5. Then $n_5 = 5k + 1$ for some integer $k \geq 0$ and $n_5|40$. It follows that $n_5 = 1$. Thus G has unique Sylow 5-subgroup H of order 5, which is normal in G. Hence G is not simple.

Example: We show that no group of order 96 is simple. Let G be a group of order $96 = 2^5 \times 3$. Denote the number of Sylow 2-subgroups by n_2. Then $n_2 = 2k + 1$ for some integer $k \geq 0$ and $n_2|96$. It follows that $n_2 = 1$ or 3. If $n_2 = 1$, then G has unique Sylow 2-subgroup H which is normal in G. Now, let $n_2 = 3$ and A, B, C be the three Sylow 2-subgroups of G. Then $|AB| = \frac{|A||B|}{|A \cap B|}$ implies that $\frac{32 \times 32}{|A \cap B|} \leq 96$ and hence $|A \cap B| \geq 10$. Also, by the Lagrange's theorem, $|A \cap B|$ divides $|A| = 32$. Thus $|A \cap B| = 16$. Since $[A : A \cap B] = 2$ and $[B : A \cap B] = 2$, so $A \cap B$ is normal in both A and B, which implies that $A, B \subseteq N(A \cap B)$. Thus $AB \subseteq N(A \cap B)$. Now $|AB| = \frac{|A||B|}{|A \cap B|} = 64$ implies that $|N(A \cap B)| \geq 64$. By the Lagrange's theorem, $|N(A \cap B)|$ divides 96. Therefore $|N(A \cap B)| = 96$ and $N(A \cap B) = G$. Hence $A \cap B$ is a normal subgroup of G of order 16.

Example: We show that no group of order 56 is simple. Let G be a group of order $56 = 7 \times 2^3$. Let n_7 be the number of Sylow 7-subgroups of G. Then $n_7 = 7k + 1$ for some integer $k \geq 0$ and $n_7|56$. Hence $n_7 = 1$ or 8. If $n_7 = 1$, then G has unique Sylow 7-subgroup of order 7, which is normal in G. Thus G is not simple. Let $n_7 = 8$ and $P_1, P_2, \cdots, P_8$ be the 8 Sylow 7-subgroups in G. Since Pi are distinct and $|P_i| = 7$, $P_i \cap P_j = \{e\}$ for $i \neq j$. Every nonidentity element of P_i is of order 7 and so G has $8 \times 6 = 48$ elements of order 7. If we denote the number Sylow 2-subgroups of G by n_2, then similarly, it follows that $n_2 = 1$ or 7. If $n_2 = 1$ then G has unique Sylow 2-subgroup of order 8, which is normal in G. Thus G is not simple. Let $n_2 = 7$ and $Q_1, Q_2, \cdots, Q_7$ be the 7 Sylow 2-subgroups. Since $Q_1 \neq Q_2$, $|Q_1 \cap Q_2| \leq 4$ and hence $Q_1 \cup Q_2$ contains at least $8 + 8 - 4 = 12$ elements of order 2, 4 or 8. Thus we get $|G| \geq 48 + 12 = 60$, a contradiction. Hence either $n_7 = 1$ or $n_2 = 1$, ensuring that G has a nontrivial proper normal subgroup. Therefore G is not simple. Now we recall the generalized Cayley theorem.

Theorem: Let G be a group and H be a subgroup of G. If $S = \{aH | a \in G\}$, then there is a homomorphism $\varphi : G \to A(S)$, the group of all permutations on S such that $\ker \varphi \subseteq H$.

Also recall that $\varphi : G \to A(G)$ is defined by $\varphi(g) = \tau_g$ for all $g \in G$, where $\tau_g : S \to S$ is given by $\tau_g(aH) = (ga)H$ for all $aH \in S$.

Now we have following two important consequences of the generalized Cayley Theorem.

Theorem: (Index Theorem) - Let G be a finite group. If G has a proper subgroup H such that $[G : H] = m$ and $|G| \nmid m!$ then G is not simple.

Proof: Let $S = \{aH | a \in G\}$. Defined $\tau_g : S \dashrightarrow S$ by $\tau_g(aH) = (ga)H$ for all $aH \in S$. Then $\tau_g \in A(S)$ and the mapping $\varphi : G \to A(S)$ defined by $\varphi(g) = \tau_g$ for all $g \in G$, is a homomorphism such that $\ker\varphi \subseteq H$, by the generalized Cayley theorem. From the first isomorphism theorem, $\frac{G}{\ker\varphi} \cong \varphi(G)$, a subgroup of A(S). Since $|A(S)| = m!$, so $\left|\frac{G}{\ker\varphi}\right|$ divides m!. But $|G| \nmid m!$ implies that $|\ker\varphi| > 1$. Thus $\ker\phi$ becomes a nontrivial proper normal subgroup of G, and G is not simple.

Example: We show that no group of order 12 is simple.

Let G be a group of order $12 = 3 \times 2^2$. Let H be a Sylow 2-subgroup of G. Then $|G| = 2^2 = 4$ implies that $[G : H] = 3$. Since $12 \nmid 3!$, so G is not simple, by the index theorem. Similarly we can show that no group of order 15, 21, 24, 28, 33 etc. is simple.

Theorem: Every finite simple group having a subgroup of index n, is isomorphic to a subgroup of A_n.

Proof: Let G be a finite simple group and H be a subgroup of G such that $[G : H] = n$. Denote $S = \{aH : a \in G\}$. Then $|S| = n$ which implies that $A(S) \cong S_n$. By the generalized Cayley theorem, there is a non-trivial homomorphism $\varphi : G \to S_n$. Then $\ker\varphi \neq G$. Since G is simple and $\text{Ker}\varphi$ is a normal subgroup of G so $\ker\varphi = \{e\}$ and φ is one-to-one. Thus $G \cong \varphi(G)$, a subgroup of S_n. Then $\varphi(G)$ consists of only even permutations or half even and half odd. If $\varphi(G)$ is of later type then the set of all even permutations in $\varphi(G)$ becomes a subgroup of index 2 and so a normal subgroup of $\varphi(G)$. This contradicts that G is simple. Thus $\varphi(G) \subseteq A_n$ and G is isomorphic to a subgroup of A_n.

If in the generalized Cayley theorem, we take H = {e}, then S = {{a} | a ∈ G} and τ_g is actually a left translation of G. Thus we can take $\tau_g : G \to G$ as:

$$\tau_g(a) = ga \text{ for all } a \in G,$$

and $\psi : G \to A(G)$ is a monomorphism. Therefore, G is isomorphic to a subgroup of the permutation group on G. This is the Cayley Theorem. Now we apply the Cayley Theorem to check simplicity of groups.

Theorem: Let G be a group of order 2m, where m is an odd integer. Then G has a subgroup of order m and hence G is not simple.

Proof: We have, by the Cayley's theorem, that there is a monomorphism $\psi : G \to A(G)$ given by:

$$\psi(g) = \tau_g \text{ for all } g \in G,$$

where $\tau_g : G \to G$ is defined by: $\tau_g(x) = gx$ for all x ∈ G. Since G is of even order, it contains an element of order 2, say a. Now $\tau_a(x) = ax$ and $\tau_a(ax) = a^2x = x$ implies that the permutation τ_a is a product of the transpositions of the form (x ax). It follows from the cancelation property of G that x ≠ ax for all x ∈ G. Since |G| = 2m, so τ_a is a product of such m transpositions; and hence τ_g is an odd permutation. Now define,

$$f : \psi(G) \to \{1, -1\}.$$

by

$$f(\sigma) = \begin{cases} 1 & \text{if } \sigma \text{ is an permutaion} \\ -1 & \text{if } \sigma \text{ is a odd permutation}. \end{cases}$$

for all σ ∈ ψ(G). Then f is a homomorphism. Since ψ(G) contains both odd permutation (viz. τ_a) and even permutation (viz. the identity permutation), so f is an epimorphism. Hence, by the first isomorphism theorem,

$$\psi(G)/\ker f \simeq \{-1, 1\}.$$

Then |ψ(G)/ ker f| = 2 implies that |G|/| ker f| = 2, that is, 2m/| ker f| = 2 and it follows that | ker f| = m. Thus ψ(G) contains a normal subgroup ker f of order m, and hence G contains a normal subgroup of order m. From this result it follows that no group of order 14, 18, 22 etc. is simple.

Now recall that if G is a group of order pn where p is a prime and p ≥ n, then G has a normal subgroup of order p. Thus no group of such order is simple. Hence no group of order 6, 10, 14, 15, 20, 21 etc. is simple. From all the results we have:

Theorem: Let 1 ≤ n < 60 be an integer which is not prime. Then no group of order n is simple. Now we show that the alternating group A5 is a simple group of order 60. We conclude this topic with the characterization of all simple groups of order 60.

Theorem: Let G be a group of order 60. If G has more than one Sylow 5-subgroup, then it is simple.

Proof: Let $|G| = 60$ and G has more than one Sylow 5-subgroup. Now the number n_5 of Sylow 5-subgroups is of the form $5k + 1$ for some integer $k \geq 0$ and $n_5 \mid 60$ implies that $n_5 = 6$, since $n_5 \neq 1$.

If possible, assume that G is not simple and H is a non-trivial proper normal subgroup of G. If $5 \mid |G|$, then H contains a Sylow 5-subgroup P of G. Since H is normal in G, and every Sylow 5-subgroup of G is a conjugate of P, so H contains all 6 Sylow 5-subgroups of G. Total number of elements in these 6 Sylow 5-subgroups is $1 + 6 \times 4 = 25$ which implies that $|H| \geq 25$. Since $|H| \mid 60$ and H is proper, so $|H| = 30$, this contradicts that every group of order 30 has unique Sylow 5-subgroup. Thus $5 \nmid |H|$.

If $|H| = 6$ or 12, then H has a normal Sylow subgroup which is also normal in G. Thus, we may assume, by replacing H by its normal subgroup if necessary, that $|H| = 2, 3, 4$. Then $|G/H| = 30$, 20 ore 15 which implies that G/H has a normal subgroup K/H of order 5, where K is a normal subgroup of G. Now $|K| = |K/H| \cdot |H| = 5 \cdot |H|$ implies that $5 \mid |K|$ which contradicts the preceding paragraph. Thus G is simple.

Corollary: A_5 is simple.

Proof: The subgroups < (12345) > an < (13245) > are distinct Sylow 5-subgroups of A_5 and so A_5 is simple.

Lemma: Every simple group of order 60 contains a subgroup of order 12.

Proof: Let G be a group of order $60 = 5 \times 3 \times 2^2$. Since G is simple, it has more than one Sylow 5-subgroup. In fact G has exactly six Sylow 5-subgroups which contain $6 \times 4 = 24$ elements of order 5.

Denote the number of Sylow 2-subgroups of G by n_2. Then $n_2 = 2k + 1$ for some integer $k \geq 0$ and $n_2 \mid 60$ implies that $n_2 = 1, 3, 5$ or 15.

Since G is simple $n_2 \neq 1$. Let $n_2 = 15$ and $B_1, B_2, \cdots, B_{15}$ be the 15 Sylow 2-subgroups of G. If $B_i \cap B_j = \{e\}$ for every $i \neq j$, then these 15 Sylow 2-subgroups contain $15 \times 3 = 45$ elements of order 2 or 4. Thus $|G| \geq 24 + 45$ which is a contradiction. Hence there are l, k such that $|B_l \cap B_k| = 2$. Then $B_l \cap B_k$ is normal in B_l as well as in B_k and so $B_lB_k \subseteq N(B_l \cap B_K)$. Since $|B_lB_k| = \frac{4\times4}{2} = 8$, it follows that $8 \leq |N(B_l \cap B_k)|$ which also divides 60. Also $|B_l| \mid |N(B_l \cap B_k)|$. Thus $|N(B_l \cap B_k)| = 12, 20$ or 60. Since G is simple, $N(B_l \cap B_k)| \neq 60$ and by the index theorem, $N(B_l \cap B_k)| \neq 20$. Thus $N(B_l \cap B_k)$ is a subgroup of order 12.

Let $n_2 = 3$ or 5, and P be a Sylow 2-subgroup of G. Then $n_2 = [G : N(P)]$ implies that $|N(P)| \neq 4$. Since P is a subgroup of N(P), so $4 \mid |N(P)|$. Also $|N(P)| \mid 60$. Thus $|N(P)| = 12, 20$ or 60. Similarly as above, we get $|N(P)| = 12$. Thus G must have a subgroup of order 12.

Theorem: If G is a simple group of order 60, then $G \simeq A_5$.

Proof: Let G be a simple group of order 60. Then G has a subgroup H of order 12. Hence, by the generalized Cayley theorem, there is a non-trivial homomorphism.

$$\psi : G \to S_5.$$

Since G is simple, $\ker\psi = \{e\}$ and ψ is one-to-one. Thus $G \simeq \psi(G)$. Then $\psi(G) \subseteq A_5$ or half of the elements of $\psi(G)$ even and half odd. In the later case the even permutation of $\psi(G)$ is a normal subgroup of $\psi(G)$ which contradicts that G is simple. Thus $\psi(G) \subseteq A_5$. Now $|\psi(G)| = |G| = 60 = |A_5|$ implies that $\psi(G) = A_5$. Hence $G \simeq A_5$.

Jordan-Hölder Theorem

The Jordan–Hölder theorem is a theorem about composition series of finite groups. A composition series is a chain of subgroups,

$$1 = H_0 \triangleleft H_1 \triangleleft H_2 \triangleleft \cdots \triangleleft H_{k-1} \triangleleft H_k = G,$$

where H_i is a maximal proper normal subgroup of H_{i+1}. By the third isomorphism theorem, this is equivalent to the statement that the quotient H_{i+1}/H_i is a simple group. This quotient is called a composition factor.

It is not hard to show that every finite group G has a composition series. The Jordan–Hölder theorem states that any two composition series of the same group have the same length and the same composition factors (up to permutation).

Example: The cyclic group $\mathbb{Z}_6$ has two composition series:

$$1 \triangleleft \mathbb{Z}_3 \triangleleft \mathbb{Z}_6 \text{ and } 1 \triangleleft \mathbb{Z}_2 \triangleleft \mathbb{Z}_6.$$

Here, $\mathbb{Z}_3$ is isomorphic to the subgroup {0, 2, 4} of $\mathbb{Z}_6$, and $\mathbb{Z}_2$ is isomorphic to the subgroup {0, 3}. Note that both composition series have the same length, and both have the same composition factors, but in a different order: $\mathbb{Z}_6 / \mathbb{Z}_3 \cong \mathbb{Z}_2$ and $\mathbb{Z}_6 / \mathbb{Z}_2 \cong \mathbb{Z}_3$.

Here is a discussion of two assertions made in the introduction: that every finite group has a composition series, and that H_i being a maximal normal subgroup of H_{i+1} is equivalent to saying that H_{i+1}/H_i is simple.

It is clear that every finite group has a composition series. To show that a maximal normal subgroup exists, the trivial subgroup is normal in every group, and if it is not maximal, there must be a larger normal subgroup containing it. Continue looking for larger normal subgroups until finding one that is maximal. The process is finite because the group is finite. The maximal normal subgroup H_{k-1}. Run the same process on subgroups of H_{k-1}.

The third isomorphism theorem states that normal subgroups of G/N are in one-to-one correspondence with normal subgroups of G containing N, via the natural correspondence coming from the standard homomorphism π: $G \to G/N$. So H_i is maximal in H_{i+1} if and only if there is no normal subgroup strictly containing H_i and strictly contained in H_{i+1}, which corresponds to there being no normal subgroup in H_{i+1}/H_i which is strictly bigger than {1} and strictly smaller than the whole quotient. This is the same as saying that H_{i+1}/H_i is simple.

Normality is not a transitive relation. That is, if K is normal in H and H is normal in G, it is not necessarily true that K is normal in G. Each of the subgroups in a composition series is normal in the succeeding one, but not necessarily normal in G itself. For example, S_4 has a composition series,

$$1 \lhd \mathbb{Z}_2 \lhd V_4 \lhd A_4 \lhd S_4.$$

but Z_2, the subgroup generated by a double transposition, is not normal in S_4.

Formal Statement

Theorem: Let G be a finite group. Consider two composition series,

$$1 = H_0 \lhd H_1 \lhd H_2 \lhd \cdots \lhd H_{k-1} \lhd H_k = G$$
$$1 = K_0 \lhd K_1 \lhd K_2 \lhd \cdots \lhd K_{l-1} \lhd K_l = G.$$

Then k = l and the list of composition factors is unique up to permutation; that is, the lists $\{H_{i+1}/H_i\}$ and $\{K_{j+1}/K_j\}$ are the same, after rearranging one of the lists suitably.

Applications

Unique factorization: The Jordan–Hölder theorem can be viewed as a generalization of the fundamental theorem of arithmetic that every integer can be factored as a product of prime numbers, essentially uniquely (up to permutation of the factors).

In fact, it is not hard to show that the fundamental theorem of arithmetic follows from Jordan–Hölder: consider the cyclic group Z_n. Every subgroup is normal, and a maximal subgroup in a cyclic group is precisely one which has prime index. (This is because the quotient, being abelian, is simple if and only if it contains no nontrivial proper subgroups, which is only true if the quotient has prime order.) So a composition series for Z_n must have the form,

$$1 \lhd \mathbb{Z}_{p_1} \lhd \mathbb{Z}_{p_1 p_2} \lhd \cdots \lhd \mathbb{Z}_{p_1 p_2 \cdots p_k} = \mathbb{Z}_n,$$

where the pi are primes. Since Z_n has a composition series, n can be factored as a product of primes, and by Jordan–Hölder, a different factorization of n leads to a different composition series,

$$1 \lhd \mathbb{Z}_{q_1} \lhd \mathbb{Z}_{q_1 q_2} \lhd \cdots \lhd \mathbb{Z}_{q_1 q_2 \cdots q_l} = \mathbb{Z}_n,$$

which must have the same composition factors up to permutation (and k = l). But the orders of the composition factors are just the primes p_i (and q_j), so these lists are the same up to permutation, which is precisely the statement of the fundamental theorem of arithmetic.

- Normal subgroups of the symmetric group: For n ≥ 5, the alternating group A_n is simple, so there is a composition series $A_n \lhd S_n$. Since the composition factor sizes are 2 and $\frac{n!}{2}$, this implies that the only other possible nontrivial normal subgroup of S_n has order 2, which can easily be ruled out by a simple analysis of the nontrivial element of the subgroup. So, for n ≥ 5, the only nontrivial proper normal subgroup of S_n is A_n. This is related to solvability of polynomial equations by radicals.

- Algebraic geometry: A slight generalization of Jordan–Hölder (to modules) allows for an elegant and robust definition of the multiplicity of the intersection of two algebraic curves. This is used in Bezout's theorem, which predicts the number of intersection points of two projective curves. The details of the definition are beyond the scope, but the idea is that the intersection multiplicity is defined as the length of a certain module, which is defined to be the length of its composition series. The composition series are not unique, but they all have the same number of terms, thanks to Jordan–Hölder.

Proof of the Theorem

This proof is fairly technical. It will help to compare with the proof of the fundamental theorem of arithmetic, and to understand the second isomorphism theorem. As with the fundamental theorem of arithmetic, the proof proceeds by induction, on $|G|$. The base case $|G| = 1$ is trivial. Now suppose the theorem has been proven for all groups strictly smaller than G.

Take two composition series $(H_1, H_2, \ldots, H_K)$ and $(K_1, K_2, \ldots, K_l)$ or G. The theorem is true for $H = H_{k-1}$ and $K = K_{l-1}$. If H = K, then we are done, as the composition series must be rearrangements of each other. If $H \neq K$, let $L = H \cap K$. Then L has a composition series consisting of groups L_j, by the inductive hypothesis. Then there are two composition series for H, the one involving the H_i and the following one:

$$1 \lhd L_1 \lhd L_2 \lhd \ldots \lhd L_{t\ 1} \lhd L \lhd H.$$

($L = H \cap K$ is a maximal subgroup of H because $H/(H \cap K) \cong G/K$, by the second isomorphism theorem)).

By induction, this composition series must be a rearrangement of the other one:

$$\left(H_1/H_0, H_2/H_1, \ldots, H/H_{k-2}\right) \sim \left(L_1/L_0, L_2/L_1, \ldots, L/L_{t-1}, H/L\right),$$

where ~ means "is the same up to permutation." Note that the lengths being the same implies that $t + 1 = k$.

Similarly we get two composition series for K, using the same L_i series for the second one. That is,

$$\left(K_1/K_0, K_2/K_1, \ldots, K/K_{l-2}\right) \sim \left(L_1/L_0, L_2/L_1, \ldots, L/L_{t-1}, K/L\right),$$

So $t + 1 = l$. This proves that $k = l$.

Now append G/H to the first pair of lists, and append G/K to the second pair of lists. This gives,

$$\left(H_1/H_0, H_2/H_1, \ldots, H/H_{k-2,} G/H\right) \sim \left(L_1/L_0, L_2/L_1, \ldots, L/L_{t-1}, H/L, G/H\right),$$

$$\left(L_1/L_0, L_2/L_1, \ldots, L/L_{t-1}, K/L, G/K\right) \sim \left(K_1/K_0, K_2/K_1, \ldots, K/K_{l-2,} G/H\right),$$

We want that the outer two lists are the same up to permutation. The inner two lists are the same except for the last two entries. But (H/L, G/H) and (K/L, G/K) are the same as (G/K, G/H) and (G/H, G/K), again by the second isomorphism theorem. So the inner two lists are the same up to permutation (a transposition of the last two factors), and the result follows.

3

Linear Algebra

The branch of algebra concerning with linear equations and linear maps is called linear algebra. Linear Algebra is used for representing geometry and basic objects. The topics elaborated in this chapter will help in gaining a better perspective about linear algebra.

Linear algebra is an area of study in mathematics that concerns itself primarily with the study of vector spaces and the linear transformations between them. Linear algebra initially emerged as a method for solving systems of linear equations. Problems like the following show up throughout all forms of mathematics, science, and engineering, giving linear algebra a very broad spectrum of use:

Example: Solve the following system of linear equations for x, y, and z:

$$x + 2y + z = 12$$
$$2x - y + z = 1$$
$$x + y - 3z = -4.$$

While there are many ways to solve these types of systems, one of special interest is by treating the three lines as a linear transformation and looking at its corresponding matrix:

$$\begin{matrix} x + 2y + z = 12 \\ 2x - y + z = 1 \\ x + y - 3z = -4 \end{matrix} \Rightarrow \begin{pmatrix} 1 & 2 & 1 \\ 2 & -1 & 1 \\ 1 & 1 & -3 \end{pmatrix} \begin{pmatrix} x \\ y \\ z \end{pmatrix} = \begin{pmatrix} 12 \\ 1 \\ -4 \end{pmatrix}.$$

Then, $\begin{pmatrix} x \\ y \\ z \end{pmatrix}$ is an element of the vector space $\mathbb{R}^3$, and the matrix describes a linear transformation from R^3 to itself. Finding the matrix's inverse then yields the answer:

$$\begin{pmatrix} \frac{2}{19} & \frac{7}{19} & \frac{3}{19} \\ \frac{7}{19} & -\frac{4}{19} & \frac{1}{19} \\ \frac{3}{19} & \frac{1}{19} & -\frac{5}{19} \end{pmatrix} \cdot \begin{pmatrix} 1 & 2 & 1 \\ 2 & -1 & 1 \\ 1 & 1 & -3 \end{pmatrix} \begin{pmatrix} x \\ y \\ z \end{pmatrix} = \begin{pmatrix} \frac{2}{19} & \frac{7}{19} & \frac{3}{19} \\ \frac{7}{19} & -\frac{4}{19} & \frac{1}{19} \\ \frac{3}{19} & \frac{1}{19} & -\frac{5}{19} \end{pmatrix} \cdot \begin{pmatrix} 12 \\ 1 \\ -4 \end{pmatrix}$$

$$\Rightarrow \begin{pmatrix} x \\ y \\ z \end{pmatrix} = \begin{pmatrix} 1 \\ 4 \\ 3 \end{pmatrix}.$$

However, linear algebra extends beyond just manipulating equations. Linear equations are commonly used to "linearize" other mathematical objects by reducing or expanding the object into a space with linear properties.

Linear algebra is used to solve systems of equations, create linear models, and approximate solutions to equations. Many areas of science and engineering require numerical methods, and linear algebra provides the vast majority of these numerical tools. One could even argue there are no technical subjects that do not benefit from linear algebra. Some fields where linear algebra is especially crucial are data science, structural engineering, machine learning, image processing, linear programming, fluid mechanics, control theory, network flow, and engineering design. In mathematics, linear algebraic methods are used throughout all areas of algebra and many areas of analysis, with far-reaching examples across the theory of modules, representation theory, Fourier analysis, and functional analysis.

The Idea of Linearity

Before the advent of computers, one of the principal roles of scientists and applied mathematicians was to find better approximations for expressions like sin x. Finding the output of sin x with, for instance, x = 7295 would require a trip to the library to consult a table of values and would not be calculated out to as many digits as is done today. Students were taught many tricks and algorithms to evaluate non-linear functions like $\sqrt{x}$, since there was no tool (calculator or computer) to do it for them.

Today many students learn to perform elementary arithmetic operations. It is reasonable that today's student could do anything a common calculator can do. However, students are less likely to be able to calculate operations outside basic ones like addition and multiplication: for instance, calculating $\sqrt{x}$ or values of sin x. The point of this story, then, is twofold:

- Students are now taught to think linearly (i.e., primarily in terms of the basic arithmetic operations), and because of this, they are much more prepared to study linear algebra than if, say, the operation sin x was given as much attention as x + y. Addition is treated as more important than the sine function.
- These linear tools are all that is necessary. Calculators and computers use methods deriving from linear algebra in order to approximate values of $\sqrt{x}$ or sin x. Computers think linearly too.

The vector space defined by several vectors is the set of everything one could get by performing the usual operations of arithmetic on them (here, this means scalar multiplication and vector addition). Intuitively, this fits hand in hand with the picture of a student combining numbers (or vectors) in any way they can.

Example: A common method for calculating (or approximating) sin x comes from the Gram-Schmidt process. It provides a method for answering questions like this: find the quintic polynomial p(x) that best approximates sin x in the region [π, 2π] in terms of minimizing the "least squares" error.

The linear algebraist dissects this example by realizing that the vector space in question is the space of all real-valued polynomials of degree at most 5, and it is equipped with the L^2 norm. The question then becomes a direct application of the Gram-Schmidt process.

Bases

Up to this point, the analogy between vector spaces and coordinate systems has not been fully explained. In a coordinate system, there is a dimension and an associated number of coordinates. In a vector space, there can be an analogous concept of dimension, but first two questions need to be asked:

- Do there exist vector spaces whose elements cannot be expressed in terms of coordinates?
- What exactly is meant by "coordinate"?

So there should only be one possible way to express any vector in terms of coordinates, and every vector should have a coordinate representation. Furthermore, the representations should be finite. In a vector space, sets of coordinates that satisfy these constraints must span the space and be linearly independent. Such a set is called a basis, and the plural of "basis" is "bases" (pron. bay-sees).

A set $\mathbb{S}$ of elements in a vector space is known as linearly independent if there is no linear combination $a_1 \cdot v_1 + a_2 \cdot v_2 + \cdots + a_k \cdot v_k = 0$, where $v_i \in \mathbb{S}$, a_i is an element of the underlying field, and $a_i \neq 0$ for all i (k is a positive integer, necessarily finite). Equivalently, a set of elements is linearly independent if and only if none of its elements can be expressed as a sum of multiples of the rest. Note that this immediately implies 0 cannot be an element of any linearly independent set.

A set $\mathbb{S}$ of elements in a vector space spans the space (or, is known as a spanning set of the vector space) if any element v of the vector space can be expressed as $v = a_1 \cdot v_1 + a_2 \cdot v_2 + \cdots + a_k \cdot v_k$ for some vectors $v_i \in \mathbb{S}$ and scalars a_i of the underlying field, for all $i \leq k$ (k is a positive integer).

A basis may be defined, equivalently, in any of the following ways:

- A basis is a set of vectors that is linearly independent and spans the space.
- A basis is a maximal linearly independent set. That is, no other element of the vector space can be added to the set to create a different linearly independent set.
- A basis is a minimal spanning set. That is, no element of the set can be removed to create another set whose elements span the vector space.

All nontrivial vector spaces have many bases, and every basis of a given vector space must have the same number of elements. For instance, $\mathbb{R}^3$ has many bases: one is the normal one in terms of coordinates, given by (1, 0, 0), (0, 1, 0) and (0, 0, 1), and another (in terms of the same coordinates) is (1, 2, 1), (2, −1, 1), and (1, 1, −1). In general, a finite set of vectors is a basis if the matrix created by placing them (their coordinate representations) in rows is square and invertible.

Definition: A vector space V is said to have dimension k if it has a basis with k elements. Since all bases in a vector space have the same number of elements, the concept of dimension is well-defined. (That is, a vector space only has one dimension).

Linear Transformations

In any area of mathematics, one of the crucial questions to ask about the central objects of study is how they interact with each other. Here, vector spaces are the central objects of study, and linear transformations are the mappings between them.

A linear transformation T from vector space V to vector space W, both over a field $\mathbb{F}$, is a function $T : V \to W$ that satisfies for any $u, v \in V$ and $\alpha \in \mathbb{F}$,

- $T(u+v)=T(u)+T(v)$,
- $T(\alpha v)=\alpha T(v)$.

A common setting in which linear transformations arise is that of the coordinate plane $\mathbb{R}^n$. A linear transformation $T : \mathbb{R}^n \to \mathbb{R}^n$ can be seen as a rotation and dilation/contraction of the plane. In general, linear transformations from a vector space V to itself are known as endomorphisms of V.

Example: A linear transformation $T : \mathbb{R}^n \to \mathbb{R}^k$ must take the form of a $k \times n$ matrix.

Consider the images of (1, 0,..., 0), (0, 1,..., 0), through (0, 0,..., 1) under the linear transformation T. Since each element of $\mathbb{R}^n$ is expressible uniquely as a linear combination of those coordinates, the image of each element is likewise expressible as a linear combination of the image of those coordinates. Thus, the $k \times n$ matrix M formed by placing together n column vectors of length k (corresponding to the images of the coordinates) satisfies for any column vector v in $\mathbb{R}^n$,

$$T(v)=Mv.$$

Uses in Mathematics

The methodology employed by linear algebra may be best described with the following reductive assessment:

- Reduce some aspect of the mathematical object to a collection of linear relationships.
- Be sure to specify the vector space in question: What is its underlying field? What is its dimension or one of its bases?
- Use theorems from linear algebra to answer questions about the object.

Such methodology is now widespread throughout different areas of mathematics, but much of this style of thinking can be traced to the origins of linear algebra. Much of its success should be attributed to the widespread use of linear relations throughout mathematics. Here is a famous problem from combinatorics that can be solved by knowing that the rank of a matrix is no greater than the rank of one of its factors.

Example: There are n citizens living in Odd town. Their main occupation was forming various clubs, which at some point started threatening the very survival of the city. In order to limit the number of clubs, the city council decreed the following innocent-looking rules:

- Each club must have an odd number of members.

- Every two clubs must have an even number of members in common.

What is the maximum number of clubs oddtown could possibly have?

On first reading the question, one might observe that (in some sense) there are 2^n possible clubs, one for each possible group of citizens. However, this does not fully take into account the additional restrictions. In order to do that, it is helpful to restate the rules in a more mathematical manner.

There are n elements in a basis. Their main occupation was forming various vectors, which at some point started threatening the very survival of the vector space. In order to limit the number of vectors, the vector space council decreed the following rules:

- Each vector must have an odd magnitude squared (dot product with itself).
- Every two (distinct) vectors must have an even dot product.

Now that the problem has been restated mathematically, what does it actually mean? What is the underlying field, and what does it mean to be a basis? Since the rules have been restated in a way such that only the parity of a vector's elements is relevant, the problem is reduced to one of modular arithmetic, and the underlying field is $\mathbb{Z}/2\mathbb{Z} \cong F_2$, the field of two elements.

Now, suppose there are mm vectors (clubs) in Oddtown, and consider the m × n matrix A made by placing the m vectors on top of each other. Then, $A_{ij} = 1$ if the ith club contains the jth citizen. Then, the rules imposed by the council appear when considering AA^T : it is an m \times m × m matrix with 1's along the diagonal and nowhere else. Thus, from linear algebra, $AA^T = Id_m$ has rank no larger than the minimum of the ranks of its factors, so m ≤ rank(A)=n. The maximum number of clubs Oddtown could possibly have is n.

Example: Functional analysis is largely concerned with building a metric/topological structure on top of an infinite vector space. These vector spaces are generally called function spaces and are built upon sets such as $\mathbb{C}((a, b), \mathbb{R})$, the space of all continuous real-valued functions on some open interval (a, b). This set can be seen as a vector space over the field $\mathbb{R}$, and a norm can be defined on it to add topological properties. The core of functional analysis focuses on the so-called L^p-spaces, which deal with certain classes of integrable functions.

Vector Space

A vector space **V** is a set that is closed under finite vector addition and scalar multiplication. The basic example is **n**-dimensional Euclidean space $\mathbb{R}^n$, where every element is represented by a list of **n** real numbers, scalars are real numbers, addition is component wise, and scalar multiplication is multiplication on each term separately.

For a general vector space, the scalars are members of a field **F**, in which case **V** is called a vector space over **F**. Euclidean **n**-space $\mathbb{R}^n$ is called a real vector space, and $\mathbb{C}^n$ is called a complex vector space.

In order for **V** to be a vector space, the following conditions must hold for all elements **X**, **Y**, **Z** ∈ **V** and any scalars r, **s** ∈ in **F**:

- Commutativity:

  $\mathbf{X}+\mathbf{Y}=\mathbf{Y}+\mathbf{X}.$

- Associativity of vector addition:

  $(\mathbf{X}+\mathbf{Y})+\mathbf{Z}=\mathbf{X}+(\mathbf{Y}+\mathbf{Z}).$

- Additive identity: For all **X**,

  $\mathbf{0}+\mathbf{X}=\mathbf{X}+\mathbf{0}=\mathbf{X}.$

- Existence of additive inverse: For any **X**, there exists a -**X** such that,

  $\mathbf{X}+(-\mathbf{X})=\mathbf{0}.$

- Associativity of scalar multiplication:

  $r(s\mathbf{X})=(rs)\mathbf{X}.$

- Distributivity of scalar sums:

  $(r+s)\mathbf{X}=r\mathbf{X}+s\mathbf{X}.$

- Distributivity of vector sums:

  $r(\mathbf{X}+\mathbf{Y})=r\mathbf{X}+r\mathbf{Y}.$

- Scalar multiplication identity:

  $1\mathbf{X}=\mathbf{X}.$

Let **V** be a vector space of dimension **n** over the field of **q** elements (where **q** is necessarily a power of a prime number). Then the number of distinct nonsingular linear operators on **V** is,

$$\begin{aligned}M(n,q)&=(q^n-q^0)(q^n-q^1)(q^n-q^2)\ldots(q^n-q^{n-1})\\&=q^{n^2}(q^{-n};q)_n\,.\end{aligned}$$

and the number of distinct **k**-dimensional subspaces of **V** is,

$$\begin{aligned}S(k,n,q)&=\frac{(q^n-q^0)(q^n-q^1)(q^n-q^2)\ldots(q^n-q^{k-1})}{M(k,q)}\\&=\frac{(q^n-1)(q^{n-1}-1)(q^{n-2}-1)\ldots q^{n-k+1}-1}{(q^k-1)(q^{k-1}-1)(q^{k-2}-1)\ldots(q-1)}\\&=\frac{q^{(k-n)n}(q^{-n};q)_k}{(q^{-n},q)_n},\end{aligned}$$

where $(\mathbf{q};\mathbf{a})_n$ is a q-Pochhammer symbol.

A consequence of the axiom of choice is that every vector space has a vector basis. A module is abstractly similar to a vector space, but it uses a ring to define coefficients instead of the field used for vector spaces. Modules have coefficients in much more general algebraic objects.

Linear Combinations

Linear combinations are obtained by multiplying matrices by scalars, and by adding them together. Let us start by giving a formal definition of linear combination.

Definition: Let $A_1,...,A_n$ be n matrices having dimension $K \times L$. A $K \times L$ matrix B is a linear combination of $A_1,...,A_n$ if and only if there exist n scalars $\alpha_1,...,\alpha_n$ called coefficients of the linear combination, such that,

$$B = \alpha_1 A_1 + ... + \alpha_n A_n.$$

In other words, if you take a set of matrices, you multiply each of them by a scalar, and you add together all the products thus obtained, then you obtain a linear combination. All the matrices involved in a linear combination need to have the same dimension (otherwise matrix addition would not be possible).

Example: Let A_1 and A_2 be 2×2 matrices defined as follows:

$$A_1 = \begin{bmatrix} 0 & 2 \\ 1 & 1 \end{bmatrix}$$

$$A_2 = \begin{bmatrix} 3 & 0 \\ 1 & 1 \end{bmatrix}$$

Let $\alpha_1 = 2$ and $\alpha_2 = -1$ be two scalars. Then, the matrix,

$$B = \alpha_1 A_1 + \alpha_2 A_2$$

is a linear combination of A_1 and A_2. It is computed as follows:

$$\begin{aligned} B &= \alpha_1 A_1 + \alpha_2 A_2 \\ &= 2 \cdot \begin{bmatrix} 0 & 2 \\ 1 & 1 \end{bmatrix} - 1 \cdot \begin{bmatrix} 3 & 0 \\ 1 & 1 \end{bmatrix} \\ &= \begin{bmatrix} 2 \cdot 0 & 2 \cdot 2 \\ 2 \cdot 1 & 2 \cdot 1 \end{bmatrix} + \begin{bmatrix} -1 \cdot 3 & -1 \cdot 0 \\ -1 \cdot 1 & -1 \cdot 1 \end{bmatrix} \\ &= \begin{bmatrix} 0 & 4 \\ 2 & 2 \end{bmatrix} + \begin{bmatrix} -3 & 0 \\ -1 & -1 \end{bmatrix} \end{aligned}$$

$$=\begin{bmatrix} 0-3 & 4+0 \\ 2-1 & 2-1 \end{bmatrix}$$

$$=\begin{bmatrix} -3 & 4 \\ 1 & 1 \end{bmatrix}.$$

Linear Combinations of Vectors

Most of the times, in linear algebra we deal with linear combinations of column vectors (or row vectors), that is, matrices that have only one column (or only one row).

Example: Let A_1, A_2 and A_3 be 3 × 1 column vectors defined as follows:

$$A_1 = \begin{bmatrix} 0 \\ 1 \\ 2 \end{bmatrix} \quad A_2 = \begin{bmatrix} 1 \\ 0 \\ -1 \end{bmatrix} \quad A_3 = \begin{bmatrix} 1 \\ 0 \\ 0 \end{bmatrix}.$$

Let B be another column vector defined as,

$$B = \begin{bmatrix} 1 \\ 1 \\ 1 \end{bmatrix}.$$

Is B a linear combination of A_1, A_2 and A_3? In order to answer this question, note that a linear combination of A_1, A_2 and A_3 with coefficients α_1, α_2 and α_3 has the following form:

$$\alpha_1 A_1 + \alpha_2 A_2 + \alpha_3 A_3 = \alpha_1 \cdot \begin{bmatrix} 0 \\ 1 \\ 2 \end{bmatrix} + \alpha_2 \cdot \begin{bmatrix} 1 \\ 0 \\ -1 \end{bmatrix} + \alpha_3 \cdot \begin{bmatrix} 1 \\ 0 \\ 0 \end{bmatrix}$$

$$= \begin{bmatrix} \alpha_1 \cdot 0 + \alpha_2 \cdot 1 + \alpha_3 \cdot 1 \\ \alpha_1 \cdot 1 + \alpha_2 \cdot 0 + \alpha_3 \cdot 0 \\ \alpha_1 \cdot 2 + \alpha_2 \cdot (-1) + \alpha_3 \cdot 0 \end{bmatrix}$$

$$= \begin{bmatrix} \alpha_2 + \alpha_3 \\ \alpha_1 \\ 2\alpha_1 - \alpha_2 \end{bmatrix}.$$

Now, B is a linear combination of A_1, A_2 and A_3 if and only if we can find α_1, α_2 and α_3 such that

$$B = \alpha_1 A_1 + \alpha_2 A_2 + \alpha_3 A_3.$$

which is equivalent to,

$$\begin{bmatrix} 1 \\ 1 \\ 1 \end{bmatrix} = \begin{bmatrix} \alpha_2 + \alpha_3 \\ \alpha_1 \\ 2\alpha_1 - \alpha_2 \end{bmatrix}$$

But we know that two vectors are equal if and only if their corresponding elements are all equal to each other. This means that the above equation is satisfied if and only the following three equations are simultaneously satisfied:

$$\begin{cases} \alpha_2 + \alpha_3 = 1 \\ \alpha_1 = 1 \\ 2\alpha_1 - \alpha_2 = 1. \end{cases}$$

The second equation gives us the value of the first coefficient:

$$\alpha_1 = 1,$$

By substituting this value in the third equation, we obtain,

$$\alpha_2 = 1,$$

Finally, by substituting the value of α_2 in the first equation, we get,

$$\alpha_3 = 0,$$

You can easily check that these values really constitute a solution to our problem:

$$1 \cdot A_1 + 1 \cdot A_2 + 0 \cdot A_3 = 1 \cdot \begin{bmatrix} 0 \\ 1 \\ 2 \end{bmatrix} + 1 \cdot \begin{bmatrix} 1 \\ 0 \\ -1 \end{bmatrix} + 0 \cdot \begin{bmatrix} 1 \\ 0 \\ 0 \end{bmatrix}$$

$$= \begin{bmatrix} 1 \cdot 0 + 1 \cdot 1 + 0 \cdot 1 \\ 1 \cdot 1 + 1 \cdot 0 + 0 \cdot 0 \\ 1 \cdot 2 + 0 \cdot (-1) + 0 \cdot 0 \end{bmatrix}$$

$$= \begin{bmatrix} 1 \\ 1 \\ 1 \end{bmatrix} = B$$

Therefore, the answer to our question is affirmative.

Linear Independence and Dependence of Vectors

A finite nonempty set of vectors $\{v_1, \ldots, v_n\}$ of V is said to be linearly independent over $\mathbb{F}$, if,

$$a_1, \ldots, a_n \in \mathbb{F} \text{ and } a_1 v_1 + \cdots + a_n v_n = 0$$

implies that $a_1 = \ldots = a_n = 0$.

Note that $\mathbf{o} = 0.v_1 + 0.v_2 + \ldots + 0.v_n$ is always true. This shows that the vector $\mathbf{o}$ is a linear combination of $\{v_1, \ldots, v_n\}$ in a trivial way. Linear independence of $\{v_1, \ldots, v_n\}$ ensures that the vector **o** cannot be expressed as a linear combination of the vectors $\{v_1, \ldots, v_n\}$ in a nontrivial way.

Definition: A nonempty subset S of V is said to be linearly independent over $\mathbb{F}$ if every finite subset of S is linearly independent.

Example: The set $\{e_1, \ldots, e_n\}$ is a linearly independent subset of F^n; where $e_i = (0, 0, \ldots, 0, 1, 0, \ldots, 0) \in F^n$, with 1 at the i-th position.

Example: The set,

$$\left\{\begin{pmatrix}1 & 0\\ 0 & 0\end{pmatrix}, \begin{pmatrix}0 & 1\\ 0 & 0\end{pmatrix}, \begin{pmatrix}0 & 0\\ 1 & 0\end{pmatrix}, \begin{pmatrix}0 & 0\\ 0 & 1\end{pmatrix}\right\}$$

is a linearly independent subset of $M_2(\mathbb{R})$.

Definition: A nonempty subset S of V is linearly dependent over $\mathbb{F}$, if there exist $v_1, \ldots, v_n$ in V and $a_1, \ldots, a_n \in \mathbb{F}$, with $a_i \neq 0$ for some i, such that,

$$a_1 v_1 + \cdots + a_n v_n = \mathbf{0}.$$

Example: $\{\mathbf{o}\}$ is linearly dependent, for, $a.\mathbf{o} = \mathbf{o}$ for all $a \in \mathbb{F}$.

Example: The expression,

$$1 \cdot (1, 1) - 1 \cdot (1, 0) + 1 \cdot (0, 1) = (0, 0)$$

shows that the set $\{(1, 1),(1, 0),(0, 1)\}$ in R^2 is linearly dependent over $\mathbb{R}$.

Theorem: Let S be a linearly independent subset of V, and let $v \in V \backslash S$. Then, $S \cup \{v\}$ is linearly dependent if and only if $v \in h\langle S\rangle$.

Proof: If $S \cup \{v\}$ is linearly dependent, then $\{v, v_1, \ldots, v_n\}$ is linearly dependent, for some $\{v_1, \ldots, v_n\} \subseteq S$. Then,

$$av + a_1 v_1 + \cdots + a_n v_n = \mathbf{0},$$

for some $a, a_1, \ldots, a_n \in F$ and $a \neq 0$. Therefore, a^{-1} exists in $\mathbb{F}$ and

$$v = -\frac{a_1}{a} v_1 - \frac{a_2}{a} v_2 - \cdots - \frac{a_n}{a} v_n.$$

This shows that $v \in \langle S\rangle$.

Conversely, if $v \in \langle S\rangle$, then,

$$v = a_1 v_1 + \cdots + a_n v_n.$$

for some $a_1, \ldots, a_n \in \mathbb{F}$. Therefore,

$$v - a_1 v_1 - \cdots - a_n v_n = \mathbf{0}.$$

This shows that $S \cup \{v\}$ is linearly dependent.

Bases and Dimension

A basis $\mathcal{B}$ for a vector space V is a linearly independent spanning subset of V.

- Example: The set of vectors $\{e_1, \ldots, e_n\}$ is a basis for F^n, where $e_i = (0, \ldots, 0, 1, 0, \ldots, 0)$ with 1 at the ith position. This is called the canonical basis of F^n.
- Example: $\{1, x, \ldots, x^n\}$ is a basis for the vector space $R_n[X]$ over $\mathbb{R}$.
- Example: $\{1, x, x^2, \ldots, x^n, \ldots\}$ is a basis for the vector space $R[X]$ over $\mathbb{R}$.

Theorem: Let $\mathcal{B} = \{v_1, \ldots, v_n\}$ be a nonempty finite subset of V . The following statements are equivalent:

- $\mathcal{B}$ is a basis of V.
- Every $v \in V$ can be written uniquely as, $v = a_1v_1 + \cdots + a_nv_n$, for some $a_1, \ldots, a_n \in \mathbb{F}$.

Proof: Every $v \in V$ can be written as $v = a_1v_1 + \cdots + a_nv_n$, for some $a_1, \ldots, a_n \in \mathbb{F}$, since $V = \langle \mathcal{B} \rangle$ - the span of $\mathcal{B}$. We have to prove the uniqueness of the expression.

Suppose that,

$$v = a_1v_1 + \cdots + a_nv_n = b_1v_1 + \cdots + b_nv_n,$$

such that $a_1, \ldots, a_n, b_1, \ldots, b_n \in \mathbb{F}$. Therefore,

$$(a_1 - b_1)v_1 + \cdots + (a_n - b_n)v_n = 0.$$

Since, $\{v_1, \ldots, v_n\}$ is linearly independent, we get,

$$a_1 - b_1 = a_2 - b_2 = \cdots = a_n - b_n = 0.$$

Hence, $a_1 = b_1, a_2 = b_2, \cdots, a_n = b_n$.

It is clear that $V = \langle \mathcal{B} \rangle$. We now prove that $\mathcal{B}$ is a linearly independent over $\mathbb{F}$.

Suppose that, for some $a_1, \ldots, a_n \in \mathbb{F}$,

$$a_1v_1 + \cdots + a_nv_n = \mathbf{0}.$$

We know that $\mathbf{0} = 0.v_1 + \cdots + 0.v_n$. Therefore, by the uniqueness of expression, we obtain $a_1 = 0$, $a_2 = 0, \ldots, a_n = 0$. This shows that $\mathcal{B}$ is linearly independent.

In the above theorem, we can take $\mathcal{B}$ to be any nonempty subset of V, not necessarily finite.

Definition: A vector space V is said to be finite dimensional over F if V has a finite spanning set S.

- Example: $\mathbb{F}^n$, $\mathbb{R}_n[X]$, $M_{m,n}(\mathbb{R})$ are some examples of finite dimensional vector spaces.
- Example: The vector space $\mathbb{R}[X]$ is not finite dimensional as a vector space over $\mathbb{R}$.

Linear Transformations

A linear transformation between two vector spaces **V** and **W** is a map **T** : **V** → **W** such that the following hold:

- $T(\mathbf{v}_1 + \mathbf{v}_2) = T(\mathbf{v}_1) + T(\mathbf{v}_2)$ for any vectors $\mathbf{v}_1$ and $\mathbf{v}_2$ in V.
- $T(\alpha \mathbf{v}) = \alpha T(\mathbf{v})$ for any scalar α.

A linear transformation may or may not be injective or surjective. When **V** and **W** have the same dimension, it is possible for T to be invertible, meaning there exists a T^{-1} such that $TT^{-1} = I$. It is always the case that T(0)=0. Also, a linear transformation always maps lines to lines (or to zero).

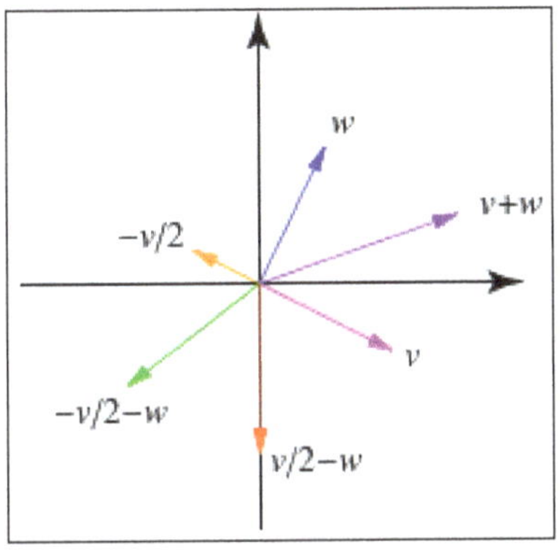

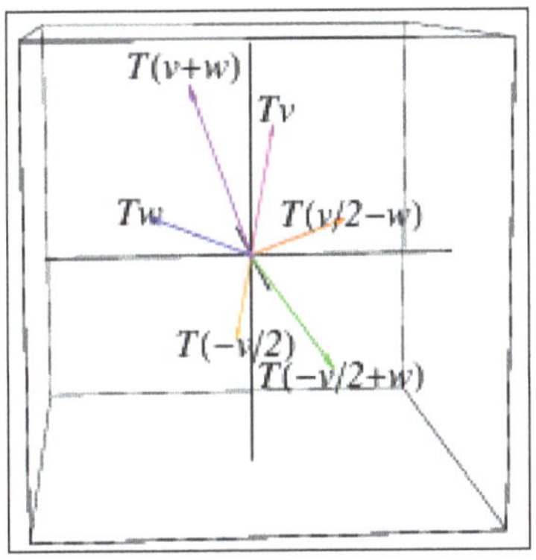

The main example of a linear transformation is given by matrix multiplication. Given an **n** × **m** matrix **A**, define T(**v**) = **Av**, where **v** is written as a column vector (with **m** coordinates). For example, consider.

$$\mathbf{A} = \begin{bmatrix} 0 & 1 \\ -2 & 2 \\ 1 & 0 \end{bmatrix},$$

then T is a linear transformation from $\mathbb{R}^2$ to $\mathbb{R}^3$, defined by,

$$T(x,y) = (y, -2x + 2y, x).$$

When **V** and **W** are finite dimensional, a general linear transformation can be written as a matrix multiplication only after specifying a vector basis for **V** and **W**. When **V** and **W** have an inner product, and their vector bases, $\{v_1,...,v_m\}$ and $\{w,...,w_n\}$, are orthonormal, it is easy to write the corresponding matrix $\mathbf{A} = (\mathbf{a}_{ij})$. In particular, $a_{ij} = \langle w_i, T(v_j)\rangle$. Note that when using the standard basis for $\mathbb{R}^n$ and $\mathbb{R}^m$, the jth column corresponds to the image of the jth standard basis vector.

When V and W are infinite dimensional, then it is possible for a linear transformation to not be continuous. For example, let V be the space of polynomials in one variable, and T be the derivative. Then $T(x^n) = n\,x^{n-1}$, which is not continuous because $x^n/n \to 0$ while $T(x^n/n)$ does not converge.

Linear two-dimensional transformations have a simple classification. Consider the two-dimensional linear transformation,

$$\rho x_1' = a_{11} x_1 + a_{12} x_2$$
$$\rho x_2' = a_{21} x_1 + a_{22} x_2.$$

Now rescale by defining $\lambda \equiv x_1 / x_2$ and $\lambda' \equiv x_1' / x_2'$. Then the above equations become,

$$\lambda' = \frac{\alpha\lambda + \beta}{\gamma\lambda + \delta},$$

where $\alpha\delta - \beta\gamma \neq 0$ and α, β, γ, and δ are defined in terms of the old constants. Solving for λ gives,

$$\lambda = \frac{\delta\lambda' - \beta}{-\gamma\lambda' + \alpha},$$

so the transformation is one-to-one. To find the fixed points of the transformation, set λ = λ' to obtain,

$$\gamma\lambda^2 + (\delta - \alpha)\lambda - \beta = 0.$$

This gives two fixed points, which may be distinct or coincident. The fixed points are classified as follows.

Variables	Type
$(\delta-\alpha)^2 + 4\beta\gamma > 0$	Hyperbolic fixed point
$(\delta-\alpha)^2 + 4\beta\gamma < 0$	Elliptic fixed point
$(\delta-\alpha)^2 + 4\beta\gamma = 0$	Parabolic fixed point

Matrix

In mathematics, a matrix (plural matrices) is a rectangular array of numbers, symbols, or expressions, arranged in rows and columns. Matrices are commonly written in box brackets. The horizontal and vertical lines of entries in a matrix are called rows and columns, respectively. The size of a matrix is defined by the number of rows and columns that it contains. A matrix with m rows and n columns is called an m × n matrix or m-by-n matrix, while m and n are called its dimensions. The dimensions of the following matrix are 2 × 3 up (read "two by three"), because there are two rows and three columns.

$$A = \begin{bmatrix} 1 & 9 & -13 \\ 20 & 5 & -6 \end{bmatrix}$$

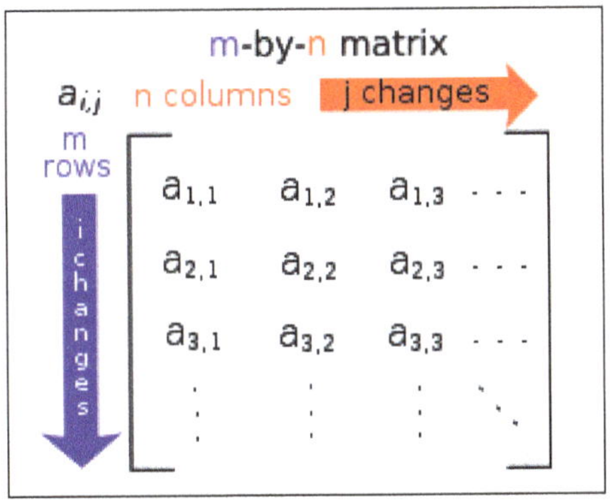

Matrix Dimensions: Each element of a matrix is often denoted by a variable with two subscripts. For instance, $a_{2,1}$ represents the element at the second row and first column of a matrix A.

The individual items (numbers, symbols or expressions) in a matrix are called its elements or entries. Provided that they are the same size (have the same number of rows and the same number of columns), two matrices can be added or subtracted element by element. The rule for matrix multiplication, however, is that two matrices can be multiplied only when the number of columns in the first equals the number of rows in the second. Any matrix can be multiplied element-wise by a scalar from its associated field.

Matrices which have a single row are called row vectors, and those which have a single column are called column vectors. A matrix which has the same number of rows and columns is called a square matrix. In some contexts, such as computer algebra programs, it is useful to consider a matrix with no rows or no columns, called an empty matrix.

Addition and Subtraction; Scalar Multiplication

Matrix addition, subtraction, and scalar multiplication are types of operations that can be applied to modify matrices. There are a number of operations that can be applied to modify matrices, such as matrix addition, subtraction, and scalar multiplication. These form the basic techniques to work with matrices.

These techniques can be used in calculating sums, differences and products of information such as sodas that come in three different flavors: apple, orange, and strawberry and two different packaging: bottle and can. Two tables summarizing the total sales between last month and this month are written to illustrate the amounts. Matrix addition, subtraction and scalar multiplication can be used to find such things as: the sales of last month and the sales of this month, the average sales for each flavor and packaging of soda in the 2-month period.

Adding and Subtracting Matrices

We use matrices to list data or to represent systems. Because the entries are numbers, we can perform operations on matrices. We add or subtract matrices by adding or subtracting corresponding entries. In order to do this, the entries must correspond. Therefore, addition and subtraction of matrices is only possible when the matrices have the same dimensions. Matrix addition is commutative and is also associative, so the following is true:

$$A + B = B + A.$$

$$(A+B)+C=A+(B+C).$$

Adding matrices is very simple. Just add each element in the first matrix to the corresponding element in the second matrix.

$$\begin{pmatrix} 1 & 2 & 3 \\ 4 & 5 & 6 \end{pmatrix} + \begin{pmatrix} 10 & 20 & 30 \\ 40 & 50 & 60 \end{pmatrix} = \begin{pmatrix} 11 & 22 & 33 \\ 44 & 55 & 66 \end{pmatrix}.$$

Note that element in the first matrix, 1, adds to element x_{11} in the second matrix, 10, to produce element x_{11} in the resultant matrix, 11. Also note that both matrices being added are 2 × 3, and the resulting matrix is also 2 × 3. You cannot add two matrices that have different dimensions.

As you might guess, subtracting works much the same way except that you subtract instead of adding.

$$\begin{pmatrix} 10 & -20 & 30 \\ 40 & 50 & 60 \end{pmatrix} - \begin{pmatrix} 1 & -2 & 3 \\ 4 & -5 & 6 \end{pmatrix} = \begin{pmatrix} 9 & -18 & 27 \\ 36 & 55 & 54 \end{pmatrix}.$$

The resulting matrix has the same dimensions as the originals, and that you cannot subtract two matrices that have different dimensions. Be careful when subtracting with signed numbers.

Scalar Multiplication

In an intuitive geometrical context, scalar multiplication of a real Euclidean vector by a positive real number multiplies the magnitude of the vector without changing its direction. What does it mean to multiply a number by 3? It means you add the number to it 3 times. Multiplying a matrix by 3 means the same thing; you add the matrix to it 3 times, or simply multiply each element by that constant.

$$3 \cdot \begin{pmatrix} 1 & 2 & 3 \\ 4 & 5 & 6 \end{pmatrix} = \begin{pmatrix} 3 & 6 & 9 \\ 12 & 15 & 18 \end{pmatrix}.$$

The resulting matrix has the same dimensions as the original. Scalar multiplication has the following properties:

- Left and right distributivity: (c + d)**M** = **M**(c + d) = **M**c + **M**d.
- Associativity: (cd)**M** = c(d**M**).
- Identity: 1**M** = M.
- Null: 0**M** = 0.
- Additive inverse: (−1)**M** = −**M**.

Matrix Multiplication

When multiplying matrices, the elements of the rows in the first matrix are multiplied with corresponding columns in the second matrix. If A is an n × m matrix and B is an m × p matrix, the result AB of their multiplication is an n × p matrix defined only if the number of columns m in A is equal to the number of rows m in B. Check to make sure that this is true before multiplying the matrices, since there is "no solution" otherwise.

Scalar multiplication is simply multiplying a value through all the elements of a matrix, whereas matrix multiplication is multiplying every element of each row of the first matrix times every element of each column in the second matrix. Scalar multiplication is much simpler than matrix multiplication; however, a pattern does exist.

When multiplying matrices, the elements of the rows in the first matrix are multiplied with corresponding columns in the second matrix. Each entry of the resultant matrix is computed one at a time. For two matrices the final position of the product is shown below:

$$\begin{bmatrix} a_{11} & a_{12} \\ \cdot & \cdot \\ a_{31} & a_{32} \\ \cdot & \cdot \end{bmatrix} \begin{bmatrix} \cdot & b_{12} & b_{13} \\ \cdot & b_{22} & b_{23} \end{bmatrix} = \begin{bmatrix} \cdot & x_{12} & \cdot \\ \cdot & \cdot & \cdot \\ \cdot & \cdot & x_{33} \\ \cdot & \cdot & \cdot \end{bmatrix}.$$

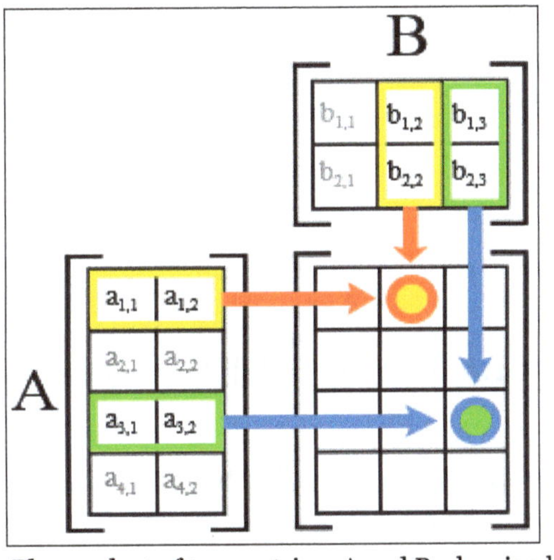

Matrix Multiplication: The product of two matrices A and B, showing how each intersection in the product matrix corresponds to a row of A and a column of B.

The values at the intersections marked with circles are:

$$x_{12} = (a_{11}, a_{12}) \cdot (b_{12}, b_{22}) = (a_{11}b_{12}) + (a_{12}b_{22})$$

$$x_{33} = (a_{31}, a_{32}) \cdot (b_{13}, b_{23}) = (a_{31}b_{13}) + (a_{32}b_{23}).$$

Matrix Multiplication: Process

Example: Find the product AB:

$$A = \begin{pmatrix} 1 & 2 \\ 3 & 4 \end{pmatrix} \quad B = \begin{pmatrix} 5 & 6 \\ 7 & 8 \end{pmatrix}.$$

First ask: Do the number of columns in A equal the number of rows in B? The number of columns in A is 2, and the number of rows in B is also 2, therefore a product exists.

Start with producing the product for the first row, first column element. Take the first row of Matrix A and multiply by the first column of Matrix B: The first element of A times the first column element of B, plus the second element of A times the second column element of B.

$$AB = \begin{pmatrix} (1\cdot 5)+(2\cdot 7) & 0+0 \\ 0+0 & 0+0 \end{pmatrix}$$

Continue the pattern with the first row of A by the second column of B, and then repeat with the second row of A.

AB has entries defined by the equation:

$$AB = \begin{pmatrix} (1\cdot 5)+(2\cdot 7) & (1\cdot 6)+(2\cdot 8) \\ (3\cdot 5)+(4\cdot 7) & (3\cdot 6)+(4\cdot 8) \end{pmatrix}$$

$$AB = \begin{pmatrix} (5+14) & (6+16) \\ (15+28) & (18+32) \end{pmatrix}$$

$$AB = \begin{pmatrix} (19) & (22) \\ (43) & (50) \end{pmatrix}.$$

The Identity Matrix

The identity matrix [I] is defined so that [A][I] = [I][A] = [A], i.e. it is the matrix version of multiplying a number by one.

The number 1 has a special property: when multiplying any number by 1, the result is the same number, i.e. 5·1 = 5. This idea can be expressed with the following property as an algebraic generalization: 1x = x. The matrix that has this property is referred to as the identity matrix.

The identity matrix, designated as [I], is defined by the property:

$$[A][I] = [I][A] = [A].$$

The definition of [I][I] stipulates that the multiplication must commute, that is, it must yield the same answer no matter in which order multiplication is done. This stipulation is important because, for most matrices, multiplication does not commute.

What matrix has this property? A first guess might be a matrix full of 1s, but that does not work:

$$\begin{pmatrix} 1 & 2 \\ 3 & 4 \end{pmatrix}\begin{pmatrix} 1 & 1 \\ 1 & 1 \end{pmatrix} = \begin{pmatrix} 3 & 3 \\ 7 & 7 \end{pmatrix},$$

So $\begin{pmatrix} 1 & 1 \\ 1 & 1 \end{pmatrix}$ is not an identity matrix.

The matrix that does work is a diagonal stretch of 1s, with all other elements being 0.

$$\begin{pmatrix} 1 & 3 \\ 2 & 4 \end{pmatrix}\begin{pmatrix} 1 & 0 \\ 0 & 1 \end{pmatrix}=\begin{pmatrix} 1 & 3 \\ 2 & 4 \end{pmatrix}$$

So $\begin{pmatrix} 1 & 0 \\ 0 & 1 \end{pmatrix}$ is the identity matrix for 2 × 2 matrices.

For a 3 × 3 matrix, the identity matrix is a 3 × 3 matrix with diagonal 1s and the rest equal to 0:

$$\begin{pmatrix} 2 & \pi & -3 \\ 5 & -2 & \frac{1}{2} \\ 9 & 8 & 8.3 \end{pmatrix}\begin{pmatrix} 1 & 0 & 0 \\ 0 & 1 & 0 \\ 0 & 0 & 1 \end{pmatrix}=\begin{pmatrix} 2 & \pi & -3 \\ 5 & -2 & \frac{1}{2} \\ 9 & 8 & 8.3 \end{pmatrix}$$

So $\begin{pmatrix} 1 & 0 & 0 \\ 0 & 1 & 0 \\ 0 & 0 & 1 \end{pmatrix}$ is the identity matrix for 3 × 3 matrices.

It is important to confirm those multiplications, and also confirm that they work in reverse order (as the definition requires).

There is no identity for a non-square matrix because of the requirement of matrices being commutative. For a non-square matrix [A] one might be able to find a matrix [I] such that [A][I]=[A], however, if the order is reversed then an illegal multiplication will be left. The reason for this is because, for two matrices to be multiplied together, the first matrix must have the same number of columns as the second has rows.

Matrix Representations

An ordered basis for the vector space V is a finite sequence of linearly independent vectors in V that generates V.

Example: Let $V = \mathbb{F}^2$. The set $\mathcal{B}_1 = \{(1, 0),(0, 1)\}$ and $\mathcal{B}_2 = \{(0, 1),(1, 0)\}$ are the same bases but different ordered bases because of the different sequencing of the basis vectors (1, 0) and (0, 1). The ordered basis $\{e_1, \ldots, e_n\}$ is called the standard basis for the vector space $\mathbb{F}^n$.

Definition: Let $\mathcal{B} = \{v_1, \ldots, v_n\}$ be an ordered basis of a finite dimensional vector space V over $\mathbb{F}$. Given $x \in V$, there exist unique $a_1, \ldots, a_n \in \mathbb{F}$, such that $x = a_1v_1 + \cdots a_nv_n$. The n-tuple $[x]_{\mathbb{B}} = (a_1, \ldots, a_n) \in \mathbb{F}^n$ is called the coordinate vector of x relative to the ordered basis $\mathcal{B}$.

Example: Let $V = M_2(\mathbb{R}) = \left\{\begin{pmatrix} a & b \\ c & d \end{pmatrix} | \, a, b, c, d \in \mathbb{R}\right\}$. An ordered basis,

for V is $\mathcal{B} = \left\{\begin{pmatrix} 1 & 0 \\ 0 & 0 \end{pmatrix}, \begin{pmatrix} 0 & 1 \\ 0 & 0 \end{pmatrix}, \begin{pmatrix} 0 & 0 \\ 1 & 0 \end{pmatrix}, \begin{pmatrix} 0 & 0 \\ 0 & 1 \end{pmatrix}\right\}$. A typical vector $x = \begin{pmatrix} a & b \\ c & d \end{pmatrix}$ can be written as

$x = \begin{pmatrix} a & b \\ c & d \end{pmatrix} = a\begin{pmatrix} 1 & 0 \\ 0 & 0 \end{pmatrix} + b\begin{pmatrix} 0 & 1 \\ 0 & 0 \end{pmatrix} + c\begin{pmatrix} 0 & 0 \\ 1 & 0 \end{pmatrix} + d\begin{pmatrix} 0 & 0 \\ 0 & 1 \end{pmatrix}$. The coordinate vector for the vector $x = \begin{pmatrix} a & b \\ c & d \end{pmatrix}$, with respect to the ordered basis $\mathcal{B}$ is $[x]_{\mathbb{B}} = (a, b, c, d)$.

Let V and W be finite dimensional vector space over $\mathbb{F}$ and $T : V \to W$ be a linear transformation. Let $B_1 = \{v_1, \ldots, v_n\}$ and $B_2 = \{w_1, \ldots, w_m\}$ be ordered bases for V and W respectively. For each $v_j \in B_1$, there exist unique scalars $a_{ij} \in \mathbb{F}$ such that,

$$T(v_j) = \sum_{i=1}^{n} a_{ij} w_i \quad \text{for} \quad 1 \le j \le n.$$

Definition: The $m \times n$ matrix $A = (a_{ij})$ is the matrix representation of T with respect to the ordered bases $\mathcal{B}_1$ and $\mathcal{B}_2$. We write $A = [T]_{\mathcal{B}_1}^{\mathcal{B}_2}$. If V = W and $\mathcal{B}_1 = \mathcal{B}_2$, we write $A = [T]_{\mathcal{B}}$.

Example: Let $T_\theta : \mathbb{R}^2 \to \mathbb{R}^2$ be the rotation operator given by,

$$T_\theta(x, y) = (x \cos\theta - y \sin\theta,\ x \sin\theta + y \cos\theta).$$

Then, $[T_\theta]_{\mathcal{B}} = \begin{pmatrix} \cos\theta & -\sin\theta \\ \sin\theta & \cos\theta \end{pmatrix}$ where $\mathbb{B}$ is the standard basis for R^2.

Example: Let $T : \mathbb{R}^2 \to \mathbb{R}^2$ be the projection map $T(x, y) = (x, 0)$. The matrix $[T]_{\mathcal{B}} = \begin{pmatrix} 1 & 0 \\ 0 & 0 \end{pmatrix}$, where $\mathbb{B}$ is the standard basis for $\mathbb{R}^2$.

Example: Let $T : \mathbb{R}^2 \to \mathbb{R}^3$ be the linear transformation given by $T(x, y) = (x, 0, y)$. Let $\mathcal{B}_1 = \{e_1, e_2\}$ and $\mathcal{B}_2 = \{e_1, e_2, e_3\}$ be the standard bases for $\mathbb{R}^2$ and $\mathbb{R}^3$ respectively. We have,

$$T(1, 0) = (1, 0, 0) = 1e_1 + 0e_2 + 0e_3$$

$$T(0, 1) = (0, 0, 1) = 0e_1 + 0e_2 + 1e_3$$

Therefore, $[T]_{\mathcal{B}_1}^{\mathcal{B}_2} = \begin{pmatrix} 1 & 0 \\ 0 & 0 \\ 0 & 1 \end{pmatrix}$.

If $\mathcal{B}_1' = \{(1, 0), (1, 1)\}$ is a different ordered basis of $\mathbb{R}^2$, then,

$$T(1, 0) = (1, 0, 0) = 1e_1 + 0e_2 + 0e_3$$

$$T(1, 1) = (0, 0, 1) = 1e_1 + 0e_2 + 1e_3$$

Therefore, $[T]_{\mathcal{B}_1'}^{\mathcal{B}_2} = \begin{pmatrix} 1 & 1 \\ 0 & 0 \\ 0 & 1 \end{pmatrix}$.

Let V and W be vector spaces over $\mathbb{F}$. Let $\mathcal{L}(V, W)$ denote the collection of all linear transformation from V to W over $\mathbb{F}$. Given $T, U \in \mathcal{L}(V, W)$ and $\alpha \in \mathbb{F}$, we define,

- $(T + U)(x) = T(x) + U(x) \ \forall x \in V;$

- $(\alpha T)(x) = \alpha T(x) \quad \forall x \in V.$

It is easy to verify that both T + U and αT are also linear transformations from V to W. The set $\mathcal{L}$(V, W) is a vector space with respect to the addition and scalar multiplication defined above. If V = W, we write $\mathcal{L}$(V) for the vector space of all linear operator T : V → V. If W = $\mathbb{F}$, the vector space $\mathcal{L}$(V, $\mathbb{F}$) is written as V* and it is called the dual space of V.

Theorem: Let V and W be finite dimensional vector spaces over $\mathbb{F}$, with ordered bases $\mathcal{B}_1$ and $\mathcal{B}_2$ respectively. Let T, U : V → W be linear transformations. Then,

- $[T+U]_{\mathcal{B}_1}^{\mathcal{B}_2} = [T]_{\mathcal{B}_1}^{\mathcal{B}_2} + [U]_{\mathcal{B}_1}^{\mathcal{B}_2}$;
- $[\alpha T]_{\mathcal{B}_1}^{\mathcal{B}_2} = \alpha [T]_{\mathcal{B}_1}^{\mathcal{B}_2}$, for all α ∈ $\mathbb{F}$.

Proof: Let $\mathcal{B}_1 = \{v_1, \ldots, v_n\}$ and $\mathcal{B}_2 = \{w_1, \ldots, w_m\}$. Let $T(v_j) = \sum_{i=1}^{m} a_{ij} w_j$ and $U(v_j) = \sum_{i=1}^{m} b_{ij} w_i$. Then,

$$(T+U)(v_j) = T(v_j) + U(v_j) = \sum_{i=1}^{m} (a_{ij} + b_{ij}) w_i .$$

and

$$(\alpha T)(v_j) = \alpha T(v_j) = \alpha \sum_{i=1}^{m} a_{ij} w_i = \sum_{i=1}^{m} (\alpha a_{ij}) w_i .$$

Composition of Linear Transformations and Matrix Multiplication

Let V_1, V_2, V_3 be vector spaces over a field $\mathbb{F}$. Let $T : V_1 \to V_2$ and $U : V_2 \to V_3$ be linear maps. Then $U \circ T : V_1 \to V_3$ is also linear, for, if x, y ∈ V_1 and α, β ∈ $\mathbb{F}$, then,

$$\begin{aligned}(U \circ T)(\alpha x + \beta y) &= U(T(\alpha x + \beta y)) \\ &= U(\alpha T(x) + \beta T(y)) \\ &= \alpha UT(x) + \beta UT(y).\end{aligned}$$

Example: Suppose that T : $\mathbb{R}^2 \to \mathbb{R}^2$ given by T(x, y) = (x, 0) and U : $\mathbb{R}^2 \to \mathbb{R}^2$ given by U(x, y) = (0, y) are the projection operators. Then, U ∘ T is the zero operator $\bar{\mathbf{0}}$ on $\mathbb{R}^2$.

Theorem: Let V_1, V_2, V_3 be finite dimensional vector spaces over a field $\mathbb{F}$, with ordered bases $\mathcal{B}_1$, $\mathcal{B}_2$, $\mathcal{B}_3$ respectively. Let $T : V_1 \to V_2$ and $U : V_2 \to V_3$ be $\mathbb{F}$-linear maps. Then,

$$[U \circ T]_{\mathcal{B}_1}^{\mathcal{B}_3} = [U]_{\mathcal{B}_2}^{\mathcal{B}_3} \cdot [T]_{\mathcal{B}_1}^{\mathcal{B}_2} .$$

On the right hand side of the above expression denotes matrix multiplication. The proof of the theorem follows from the definitions.

Example: If we take the projection operators T(x, y) = (x, 0) and U(x, y) = (0, y), then U ∘ T is the zero operator $\bar{\mathbf{0}}$ on $\mathbb{R}^2$. The matrices for T and U with respect to the standard basis $\mathcal{B}$ of $\mathbb{R}^2$ are $[T]_{\mathcal{B}} = \begin{pmatrix} 1 & 0 \\ 0 & 0 \end{pmatrix}$, and $[U]_{\mathcal{B}} = \begin{pmatrix} 0 & 0 \\ 0 & 1 \end{pmatrix}$. The product $[U]_{\mathcal{B}} \cdot [T]_{\mathcal{B}}$ is indeed the zero matrix.

Rank of Matrix

The maximum number of linearly independent rows in a matrix A is called the row rank of A, and the maximum number of linearly independent columns in A is called the column rank of A. If A is an m by n matrix, that is, if A has m rows and n columns, then it is obvious that,

$$\left.\begin{aligned} \text{row rank of } A \le m \\ \text{column rank of } A \le n \end{aligned}\right\} \ (*)$$

What is not so obvious, however, is that for any matrix A, the row rank of A = the column rank of A Because of this fact, there is no reason to distinguish between row rank and column rank; the common value is simply called the rank of the matrix. Therefore, if A is m x n, it follows from the inequalities in (*) that,

$$\operatorname{rank}\left(A_{m\times n}\right) \le \min(m, n) \quad (**)$$

where min(m, n) denotes the smaller of the two numbers m and n (or their common value if m = n). For example, the rank of a 3 x 5 matrix can be no more than 3, and the rank of a 4 x 2 matrix can be no more than 2. A 3 x 5 matrix,

$$\begin{bmatrix} * & * & * & * & * \\ * & * & * & * & * \\ * & * & * & * & * \end{bmatrix}$$

can be thought of as composed of three 5-vectors (the rows) or five 3-vectors (the columns). Although three 5-vectors could be linearly independent, it is not possible to have five 3-vectors that are independent. Any collection of more than three 3-vectors is automatically dependent. Thus, the column rank—and therefore the rank—of such a matrix can be no greater than 3. So, if A is a 3 x 5 matrix, this argument shows that,

$$\operatorname{rank}\left(A_{3\times 5}\right) \le 3 = \min(3, 5)$$

in accord with (**).

The process by which the rank of a matrix is determined can be illustrated by the following example. Suppose A is the 4 x 4 matrix,

$$\begin{bmatrix} 1 & -2 & 0 & 4 \\ 3 & 1 & 1 & 0 \\ -1 & -5 & -1 & 8 \\ 3 & 8 & 2 & -12 \end{bmatrix}$$

The four row vectors,

$$\mathbf{r}_1 = (1, -2, 0, 4)$$

$$\mathbf{r}_2 = (3, 1, 1, 0)$$
$$\mathbf{r}_3 = (-1, -5, -1, 8)$$
$$\mathbf{r}_4 = (3, 8, 2, -12)$$

are not independent, since, for example,

$$\mathbf{r}_3 = 2\mathbf{r}_1 - \mathbf{r}_2 \text{ and } \mathbf{r}_4 = -3\mathbf{r}_1 + 2\mathbf{r}_2 \quad (***)$$

The fact that the vectors $\mathbf{r}_3$ and $\mathbf{r}_4$ can be written as linear combinations of the other two ($\mathbf{r}_1$ and $\mathbf{r}_2$, which are independent) means that the maximum number of independent rows is 2. Thus, the row rank—and therefore the rank—of this matrix is 2.

The equations in (***) can be rewritten as follows:

$$-2\mathbf{r}_1 + \mathbf{r}_2 + \mathbf{r}_3 = \mathbf{0} \text{ and } 3\mathbf{r}_1 - 2\mathbf{r}_2 + \mathbf{r}_4 = \mathbf{0}$$

The first equation here implies that if −2 times that first row is added to the third and then the second row is added to the (new) third row, the third row will be become **o**, a row of zeros. The second equation above says that similar operations performed on the fourth row can produce a row of zeros there also. If after these operations are completed, −3 times the first row is then added to the second row (to clear out all entries below the entry $\mathbf{a}_{11} = 1$ in the first column), these elementary row operations reduce the original matrix A to the echelon form,

$$\begin{bmatrix} 1 & -2 & 0 & 4 \\ 0 & 7 & 1 & -12 \\ 0 & 0 & 0 & 0 \\ 0 & 0 & 0 & 0 \end{bmatrix}$$

The fact that there are exactly 2 nonzero rows in the reduced form of the matrix indicates that the maximum number of linearly independent rows is 2; hence, rank A = 2, in agreement with the conclusion above. In general, then, to compute the rank of a matrix, perform elementary row operations until the matrix is left in echelon form; the number of nonzero rows remaining in the reduced matrix is the rank. [Note: Since column rank = row rank, only two of the four columns in A— $\mathbf{c}_1$, $\mathbf{c}_2$, $\mathbf{c}_3$, and $\mathbf{c}_4$—are linearly independent. Show that this is indeed the case by verifying the relations,

$$\mathbf{c}_2 = -2\mathbf{c}_1 + 7\mathbf{c}_3 \text{ and } \mathbf{c}_4 = 4\mathbf{c}_1 - 12\mathbf{c}_3$$

(and checking that $\mathbf{c}_1$ and $\mathbf{c}_3$ are independent). The reduced form of A makes these relations especially easy to see.

Example: Find the rank of the matrix,

$$B = \begin{bmatrix} 2 & -1 & 3 \\ 1 & 0 & 1 \\ 0 & 2 & -1 \\ 1 & 1 & 4 \end{bmatrix}$$

First, because the matrix is 4 × 3, its rank can be no greater than 3. Therefore, at least one of the four rows will become a row of zeros.

Perform the following row operations:

$$\begin{bmatrix} 2 & -1 & 3 \\ 1 & 0 & 1 \\ 0 & 2 & -1 \\ 1 & 1 & 4 \end{bmatrix} \xrightarrow{r_1 \leftrightarrow r_2} \begin{bmatrix} 1 & 0 & 1 \\ 2 & -1 & 3 \\ 0 & 2 & -1 \\ 1 & 1 & 4 \end{bmatrix}$$

$$\xrightarrow[-r_1 \text{ added to } r_4]{-2r_1 \text{ added to } r_2} \begin{bmatrix} 1 & 0 & 1 \\ 0 & -1 & 1 \\ 0 & 2 & -1 \\ 1 & 1 & 3 \end{bmatrix}$$

$$\xrightarrow[r_2 \text{ added to } r_4]{2r_2 \text{ added to } r_3} \begin{bmatrix} 1 & 0 & 1 \\ 0 & -1 & 1 \\ 0 & 0 & 1 \\ 0 & 0 & 4 \end{bmatrix}$$

$$\xrightarrow[-4r_3 \text{ added to } r_4]{(-1)r_2} \begin{bmatrix} 1 & 0 & 1 \\ 0 & 1 & -1 \\ 0 & 0 & 1 \\ 0 & 0 & 0 \end{bmatrix}$$

Since there are 3 nonzero rows remaining in this echelon form of B, $\operatorname{rank} B = 3$.

Example: Determine the rank of the 4 by 4 checkerboard matrix,

$$C = \begin{bmatrix} 1 & -1 & 1 & -1 \\ -1 & 1 & -1 & 1 \\ 1 & -1 & 1 & -1 \\ -1 & 1 & -1 & 1 \end{bmatrix}$$

Since $\mathbf{r}_2 = \mathbf{r}_4 = -\mathbf{r}_1$ and $\mathbf{r}_3 = \mathbf{r}_1$, all rows but the first vanish upon row-reduction:

$$\begin{bmatrix} 1 & -1 & 1 & -1 \\ -1 & 1 & -1 & 1 \\ 1 & -1 & 1 & -1 \\ -1 & 1 & -1 & 1 \end{bmatrix} \xrightarrow[\substack{-r_1 \text{ added to } r_3 \\ r_1 \text{ added to } r_4}]{r_1 \text{ added to } r_2} \begin{bmatrix} 1 & -1 & 1 & -1 \\ 0 & 0 & 0 & 0 \\ 0 & 0 & 0 & 0 \\ 0 & 0 & 0 & 0 \end{bmatrix}$$

Since only 1 nonzero row remains, rank C = 1.

Determinant of a Square Matrix

A determinant is a real number associated with every square matrix. The determinant of a square matrix A is denoted by "det A" or | A |. Now, that last one looks like the absolute value of A, but you will have to apply context. If the vertical lines are around a matrix, it means determinant. The line below shows the two ways to write a determinant:

$$\begin{vmatrix} 3 & 1 \\ 5 & 2 \end{vmatrix} = \det\begin{bmatrix} 3 & 1 \\ 5 & 2 \end{bmatrix}$$

Determinant of a 2 × 2 Matrix

The determinant of a 2 × 2 matrix is found much like a pivot operation. It is the product of the elements on the main diagonal minus the product of the elements off the main diagonal.

$$\begin{vmatrix} a & b \\ c & d \end{vmatrix} = ad - bc$$

Properties of Determinants

- The determinant is a real number, it is not a matrix.
- The determinant can be a negative number.
- It is not associated with absolute value at all except that they both use vertical lines.
- The determinant only exists for square matrices (2 × 2, 3 × 3, ... n × n). The determinant of a 1 × 1 matrix is that single value in the determinant.
- The inverse of a matrix will exist only if the determinant is not zero.

Expansion using Minors and Cofactors

The definition of determinant that we have so far is only for a 2 × 2 matrix. There is a shortcut for a 3 × 3 matrix. The method is called expansion using minors and cofactors. Before we can use them, we need to define them.

Minors

A minor for any element is the determinant that results when the row and column that element are in are deleted. The notation M_{ij} is used to stand for the minor of the element in row i and column j. So M_{21} would mean the minor for the element in row 2, column 1. Consider the 3 × 3 determinants shown below:

$$\begin{array}{c} \\ \mathbf{R_1} \\ \mathbf{R_2} \\ \mathbf{R_3} \end{array}\begin{array}{c} \begin{array}{ccc} \mathbf{C_1} & \mathbf{C_2} & \mathbf{C_3} \end{array} \\ \begin{vmatrix} 1 & 3 & 2 \\ 4 & 1 & 3 \\ 2 & 5 & 2 \end{vmatrix} \end{array}$$

Finding the Minor for R_2C_1

The minor is the determinant that remains when you delete the row and column of the element you're trying to find the minor for. That means we should delete row 2 and column 1 and then find the determinant.

$$\begin{array}{c} \\ R_1 \\ R_3 \end{array} \begin{array}{c} \begin{matrix} C_2 & C_3 \end{matrix} \\ \begin{vmatrix} 3 & 2 \\ 5 & 2 \end{vmatrix} \end{array} = 3(2) - 5(2) = 6 - 10 = -4$$

As you can see, the minor for row 2 and column 1 is $M_{21} = -4$.

Finding the Minor for R_3C_2

This time, we would delete row 3 and column 2.

$$\begin{array}{c} \\ R_1 \\ R_2 \end{array} \begin{array}{c} \begin{matrix} C_1 & C_3 \end{matrix} \\ \begin{vmatrix} 1 & 2 \\ 4 & 3 \end{vmatrix} \end{array} = 1(3) - 4(2) = 3 - 8 = -5$$

So the minor for row 3, column 2 is $M_{32} = -5$.

Matrix of Minors

When you're just trying to find the determinant of a matrix, this is overkill. But there is one extremely useful application for it and it will give us practice finding minors. The matrix of minors is the square matrix where each element is the minor for the number in that position. Here is a generic matrix of minors for a 3 × 3 determinant.

$$\begin{array}{c} \\ R_1 \\ R_2 \\ R_3 \end{array} \begin{array}{c} \begin{matrix} C_1 & C_2 & C_3 \end{matrix} \\ \begin{bmatrix} M_{11} & M_{12} & M_{13} \\ M_{21} & M_{22} & M_{23} \\ M_{31} & M_{32} & M_{33} \end{bmatrix} \end{array}$$

Let's find the matrix of minors for our original determinant. Here is the determinant:

$$\begin{array}{c} \\ R_1 \\ R_2 \\ R_3 \end{array} \begin{array}{c} \begin{matrix} C_1 & C_2 & C_3 \end{matrix} \\ \begin{vmatrix} 1 & 3 & 2 \\ 4 & 1 & 3 \\ 2 & 5 & 2 \end{vmatrix} \end{array}$$

Here is the work to find each minor in the matrix of minors.

	C_1	C_2	C_3
R_1	$\begin{vmatrix} 1 & 3 \\ 5 & 2 \end{vmatrix} = 2-15=-13$	$\begin{vmatrix} 4 & 3 \\ 2 & 2 \end{vmatrix} = 8-6=2$	$\begin{vmatrix} 4 & 1 \\ 2 & 5 \end{vmatrix} = 20-2=18$
R_2	$\begin{vmatrix} 3 & 2 \\ 5 & 2 \end{vmatrix} = 6-10=-4$	$\begin{vmatrix} 1 & 2 \\ 2 & 2 \end{vmatrix} = 2-4=-2$	$\begin{vmatrix} 1 & 3 \\ 2 & 5 \end{vmatrix} = 5-6=-1$
R_3	$\begin{vmatrix} 3 & 2 \\ 1 & 3 \end{vmatrix} = 9-2=7$	$\begin{vmatrix} 1 & 2 \\ 4 & 3 \end{vmatrix} = 3-8=-5$	$\begin{vmatrix} 1 & 3 \\ 4 & 1 \end{vmatrix} = 1-12=-11$

Finally, here is the matrix of minors. Again, you don't need to put the labels for the row and columns on there, but it may help you.

$$\begin{array}{c} \\ R_1 \\ R_2 \\ R_3 \end{array} \begin{array}{c} \begin{array}{ccc} C_1 & C_2 & C_3 \end{array} \\ \begin{bmatrix} -13 & 2 & 18 \\ -4 & -2 & -1 \\ 7 & -5 & -11 \end{bmatrix} \end{array}$$

Cofactors

A cofactor for any element is either the minor or the opposite of the minor, depending on where the element is in the original determinant. If the row and column of the element add up to be an even number, then the cofactor is the same as the minor. If the row and column of the element add up to be an odd number, then the cofactor is the opposite of the minor.

Sign Chart

Rather than adding up the row and column of the element to see whether it is odd or even, many people prefer to use a sign chart. A sign chart is either a + or - for each element in the matrix. The first element (row 1, column 1) is always a + and it alternates from there.

The + does not mean positive and the - negative. The + means the same sign as the minor and the - means the opposite of the minor. Think of it addition and subtraction rather than positive or negative.

Here is the sign chart for a 2 × 2 determinant:

$$\begin{array}{c} \\ R_1 \\ R_2 \end{array} \begin{array}{c} \begin{array}{cc} C_1 & C_2 \end{array} \\ \begin{vmatrix} + & - \\ - & + \end{vmatrix} \end{array}$$

Here is the sign chart for a 3 × 3 determinant:

$$\begin{array}{c} \\ R_1 \\ R_2 \\ R_3 \end{array}\begin{array}{c} \begin{array}{ccc} C_1 & C_2 & C_3 \end{array} \\ \begin{vmatrix} + & - & + \\ - & + & - \\ + & - & + \end{vmatrix} \end{array}$$

Matrix of Cofactors

Again, if all you're trying to do is find the determinant, you do not need to go through this much work. The matrix of cofactors is the matrix found by replacing each element of a matrix by its cofactor. This is the matrix of minors with the signs changed on the elements in the - positions.

$$\begin{array}{c} \\ R_1 \\ R_2 \\ R_3 \end{array}\begin{array}{c} \begin{array}{ccc} C_1 & C_2 & C_3 \end{array} \\ \begin{bmatrix} -13 & -2 & 18 \\ 4 & -2 & 1 \\ 7 & 5 & -11 \end{bmatrix} \end{array}$$

Expanding to Find the Determinant

Here are the steps to go through to find the determinant:

- Pick any row or column in the matrix. It does not matter which row or which column you use, the answer will be the same for any row. There are some rows or columns that are easier than others, but we'll get to that later.
- Multiply every element in that row or column by its cofactor and add. The result is the determinant.

Let's expand our matrix along the first row:

$$\begin{vmatrix} 1 & 3 & 2 \\ 4 & 1 & 3 \\ 2 & 5 & 2 \end{vmatrix}$$

From the sign chart, we see that 1 is in a positive position, 3 is in a negative position and 2 is in a positive position. By putting the + or - in front of the element, it takes care of the sign adjustment when going from the minor to the cofactor.

$$+1\begin{vmatrix} 1 & 3 \\ 5 & 2 \end{vmatrix} -3\begin{vmatrix} 4 & 3 \\ 2 & 2 \end{vmatrix} +2\begin{vmatrix} 4 & 1 \\ 2 & 5 \end{vmatrix}$$

$= 1\,(2 - 15) - 3\,(8 - 6) + 2\,(20 - 2)$

$= 1\,(-13) - 3\,(2) + 2\,(18)$

= -13 - 6 + 36

= 17

The determinant of this matrix is 17.

It doesn't really matter which row or column you use. Let's try it again, but this time expands on the second columns. As an effort to save time, the minors for that column (from the matrix of minors) were 2, -2, and -5. The original elements were 3, 1, and 5. The 3 and 5 are in negative positions.

Determinant = - 3 (2) + 1 (-2) - 5 (-5) = -6 -2 + 25 = 17

Expand on any row or any column, you'll get 17.

However, you can't do diagonals. If we try the main diagonal, you get,

+ 1 (-13) + 1 (-2) + 2 (-11) = -13 -2 - 22 = -37

Some Rows or Columns are Better than others

- Pick the row or column with the most zeros in it. Since each minor or cofactor is multiplied by the element in the matrix, picking a row or column with lots of zeros in it means that you will be multiplying by a lot of zeros. Multiplying by zero doesn't take very long at all. In fact, if the element is zero, you don't need to even find the minor or cofactor.
- Pick the row or column with the largest numbers (or variables) in it. The elements in the row or column that you expand along are not used to find the minors. The only place that they are multiplied is once, in the expansion. If you pick the row or column with the smallest numbers, then every minor will be the product of larger numbers.
- If you pick a row or column that has variables in it, then you will only have to multiply by the variables once, during the expansion.

Inverse of a Matrix (Revisited)

Let's consider our original determinant as a matrix this time.

$$\begin{bmatrix} 1 & 3 & 2 \\ 4 & 1 & 3 \\ 2 & 5 & 2 \end{bmatrix}$$

Find the matrix of minors as explained above.

$$\begin{bmatrix} -13 & -2 & 18 \\ 4 & -2 & 1 \\ 7 & -5 & -11 \end{bmatrix}$$

Turn it into a matrix of cofactors by changing the signs on the appropriate elements based on the sign chart.

$$\begin{bmatrix} -13 & -2 & 18 \\ 4 & -2 & 1 \\ 7 & 5 & -11 \end{bmatrix}$$

Find the adjoint by transposing the matrix of cofactors.

To transpose a matrix, you switch the rows and columns. That is, the rows become columns and the columns become rows. The Transpose of a matrix can be found using the TI-82 or TI-83 calculator by entering the name of the matrix and then choosing Matrix, Math, and then option 2, a superscripted T, like $[A]^T$.

$$\begin{bmatrix} -13 & 4 & 7 \\ -2 & -2 & 5 \\ 18 & 1 & -11 \end{bmatrix}$$

Finally divide the adjoint of the matrix by the determinant of the matrix. In this problem, the determinant is 17, so we'll divide every element by 17. The resulting matrix is the inverse of the original matrix.

$$\begin{bmatrix} -13/17 & 4/17 & 7/17 \\ -2/17 & -2/17 & 5/17 \\ 18/17 & 1/17 & -11/17 \end{bmatrix}$$

The inverse of a matrix is found by dividing the adjoint of the matrix by the determinant of the matrix. Don't try that on your calculator since the calculator won't let you divide a matrix by a scalar. You will have to multiply by the inverse of the determinant instead.

If you check it with your calculator, you can verify that the inverse actually is the adjoint divided by the determinant. Since the inverse is the adjoint divided by the determinant, we can see why the inverse doesn't exist if the determinant is zero. That would cause division by zero, which is undefined.

Larger Order Determinants

Let's find the determinant of a 4 × 4 system.

	C_1	C_2	C_3	C_4
R_1	3	2	0	1
R_2	4	0	1	2
R_3	3	0	2	1
R_4	9	2	3	1

Pick the row or column with the most zeros in it. In this case, that is the second column. For each element in the original matrix, its minor will be a 3 × 3 determinant. We'll have to expand each of those by using three 2 × 2 determinants.

This is why we want to expand along the second column. The minors are multiplied by their elements, so if the element in the original matrix is 0, it doesn't really matter what the minor is and we can save a lot of time by not having to find it. In the second column, you won't need to find two of the minors because their corresponding element in the second column is zero.

$$-2\begin{vmatrix} 4 & 1 & 2 \\ 3 & 2 & 1 \\ 9 & 3 & 1 \end{vmatrix} + 0\begin{vmatrix} & ? & \end{vmatrix} - 0\begin{vmatrix} & ? & \end{vmatrix} + 2\begin{vmatrix} 3 & 0 & 1 \\ 4 & 1 & 2 \\ 3 & 2 & 1 \end{vmatrix}$$

We could actually fill in those middle two minors, but since they're multiplied by 0, it doesn't really matter what they are. In fact, you could just as easily skip them.

Now, there are two 3 × 3 determinants left to find.

In the first 3 × 3 determinant, there are no zeros, so pick the row or column with the largest numbers. That would be column 1, so expand along the first column.

Notice the 4 is in a positive position. The sign charts begin over with each new determinant. The position of the number in the original matrix does not matter, only its position in the current matrix.

$$\begin{vmatrix} 4 & 1 & 2 \\ 3 & 2 & 1 \\ 9 & 3 & 1 \end{vmatrix} = +4\begin{vmatrix} 2 & 1 \\ 3 & 1 \end{vmatrix} - 3\begin{vmatrix} 1 & 2 \\ 3 & 1 \end{vmatrix} + 9\begin{vmatrix} 1 & 2 \\ 2 & 1 \end{vmatrix}$$

= 4 (2 - 3) - 3 (1 - 6) + 9 (1 - 4) = 4 (-1) - 3 (-5) + 9 (-3) = -4 + 15 - 27 = -16

Consider the other 3 × 3 matrix. In this one, there is a 0 in the row 1 and column 2. Either one of those would be a good pick for expansion, but since row 1 has slightly larger numbers, we'll expand along the first row.

$$\begin{vmatrix} 3 & 0 & 1 \\ 4 & 1 & 2 \\ 3 & 2 & 1 \end{vmatrix} = +3\begin{vmatrix} 1 & 2 \\ 2 & 1 \end{vmatrix} - 0\begin{vmatrix} ? & ? \\ ? & ? \end{vmatrix} + 1\begin{vmatrix} 4 & 1 \\ 3 & 2 \end{vmatrix}$$

= 3 (1 - 4) - 0 (doesn't matter) + 1 (8 - 3) = 3 (-3) + 1 (5) = -9 + 5 = -4

When you go to find the determinant, remember that there were elements from the original 4 × 4 matrix that were times each of those 3 × 3 determinants. The first one was -2 and the second one was +2.

Determinant = -2 (-16) + 2 (-4) = 32 - 8 = 24

Worst Case Scenario

- To find a 3 × 3 determinant with no zeros, you have to find three 2 × 2 determinants.
- To find a 4 × 4 determinant with no zeros, you have to find four 3 × 3 determinants, each of which then becomes three 2 × 2 determinants for a total of twelve 2 × 2 determinants.

- To find a 5 × 5 determinant with no zeros, you have to find five 4 × 4 determinants, each of which then becomes four 3 × 3 determinants, each of those becoming three 2 × 2 determinants for a total of sixty 2 × 2 determinants.

Using the Calculator

After that last problem, you've got to be asking yourself if there isn't an easier way. Well, yes, there is, as long as the determinant doesn't have any variables in it. You can use the calculator. The notation that the TI-82 or TI-83 calculator uses is the Det A notation. So, after entering the matrix into one of the available matrices on the calculator, enter DET by going Matrix, Math, and choosing option 1. Then put in the name of the matrix that you're using.

You don't need to use parentheses (unless you have a TI-83), but you can if you want to find the determinant of a product "det ([A]*[B])" or the determinant of a transpose "det ([A]^T)" as opposed to the transpose of the determinant "(det [A])^T". By the way, the calculator won't find the transpose of a determinant because the determinant is a scalar (real number) and the calculator only knows how to find the transpose of a matrix. The transpose of a scalar is that scalar.

Triangular Matrices

- Upper Triangular Matrix: A matrix in which all the non-zero elements are either on or above the main diagonal. That is, all the non-zero values are in the upper triangle. Everything below the diagonal is a zero.
- Lower Triangular Matrix: A matrix in which all the non-zero elements are either on or below the main diagonal. That is, all the non-zero values are in the lower triangle. Everything above the diagonal is zero.
- Diagonal Matrix: A matrix in which all the non-zero elements are on the main diagonal. Everything off the main diagonal is a zero.

The determinant of a triangular matrix or a diagonal matrix is the product of the elements on the main diagonal.

Elementary Row Operations

There were three elementary row operations that could be performed that would return an equivalent system. With determinants, since the determinant of a transpose is the same as the determinant of the matrix, the elementary row operations can also be applied to columns.

By performing row-reduction (using pivoting on a 1 if you like), you can place a matrix into triangular form. Once it's in triangular form, then all you have to do is multiply by the elements on the main diagonal and you have the determinant.

Let's look at each of the three elementary row operations:

- If you interchange two rows or two columns in a determinant, the resulting determinant will differ only in sign. That is, if you swap rows or columns, the resulting determinant is the opposite of the original determinant.

- If you multiply a row or column by a non-zero constant, the determinant is multiplied by that same non-zero constant.
- If you multiply a row or column by a non-zero constant and add it to another row or column, replacing that row or column, there is no change in the determinant.

That last operation is equivalent to pivoting on a one. If your pivot is a number other than one, then you are multiplying each row that you change by the pivot element. So, if you pivot on a 3 and you change two rows, then the resulting determinant will be 3*3 = 9 times as great as the original determinant. You do not have to place the matrix into reduced row-echelon form or even row-echelon form. You are free to stop the reduction at any point and expand using minors and cofactors.

Determinants that are Zero

The determinant of a matrix will be zero if:

- An entire row is zero.
- Two rows or columns are equal.
- A row or column is a constant multiple of another row or column.

Remember, that a matrix is invertible, non-singular, if and only if the determinant is not zero. So, if the determinant is zero, the matrix is singular and does not have an inverse.

Diagonalization of a Linear Operator

Some linear operators $T : V \to V$ have the nice property that there is some basis for V so that the matrix representing T is a diagonal matrix. We'll call those operators diagonalizable operators. We'll call a square matrix A a diagonalizable matrix if it is conjugate to a diagonal matrix, that is, there exists an invertible matrix P so that $P^{-1}AP$ is a diagonal matrix. That's the same as saying that under a change of basis, A becomes a diagonal matrix.

Reflections are examples of diagonalizable operators as are rotations if C is your field of scalars. Not all linear operators are diagonalizable. The simplest one is $\mathbf{R}^2 \to \mathbf{R}^2$, $(x, y) \to (y, 0)$ whose matrix is $A = \begin{bmatrix} 0 & 1 \\ 0 & 0 \end{bmatrix}$. No conjugate of it is diagonal. It's an example of a nilpotent matrix, since some power of it, namely A^2, is the 0-matrix. In general, nilpotent matrices aren't diagonalizable. There are many other matrices that aren't diagonalizable as well.

Theorem: A linear operator on an n-dimensional vector space is diagonalizable if and only if it has a basis of n eigenvectors, in which case the diagonal entries are the eigenvalues for those eigenvectors.

Proof: If it's diagonalizable, then there's a basis for which the matrix representing it is diagonal. The transformation therefore acts on the ith basis vector by multiplying it by the ith diagonal entry, so it's an eigenvector. Thus, all the vectors in that basis are eigenvectors for their associated diagonal entries.

Conversely, if you have a basis of n eigenvectors, then the matrix representing the transformation is diagonal since each eigenvector is multiplied by its associated eigenvalue. We'll see soon that if a linear operator on an n- dimensional space has n distinct eigenvalues, then it's diagonalizable. But first, a preliminary theorem.

Theorem: Eigenvectors that are associated to distinct eigenvalues are independent. That is, if λ_1, λ_2, . . ., λ_k are different eigenvalues of an operator T, and an eigenvector $\mathbf{v}_i$ is associated to each eigenvalue λ_i, then the set of vectors $\mathbf{v}_1$, $\mathbf{v}_2$, . . ., $\mathbf{v}_k$ is linearly independent.

Proof: Assume by induction that the first k – 1 of the eigenvectors are independent. We'll show all k of them are. Suppose some linear combination of all k of them equals **0**:

$$c_1\mathbf{v}_1 + c_2\mathbf{v}_2 + \cdots + c_k\mathbf{v}_k = \mathbf{0}.$$

Take $T - \lambda_k I$ of both sides of that equation. The left side simplifies,

$$\begin{aligned}
&(T-\lambda_k I)(c_1\mathbf{v}_1 + \cdots + c_k\mathbf{v}_k) \\
&= c_1 T(\mathbf{v}_1) - \lambda_k c_1 \mathbf{v}_1 + \cdots + c_k T(\mathbf{v}_k) - \lambda_k c_k \mathbf{v}_k \\
&= c_1(\lambda_1 - \lambda_k)\mathbf{v}_1 + \cdots + c_k(\lambda_k - \lambda_k)\mathbf{v}_k \\
&= c_1(\lambda_1 - \lambda_k)\mathbf{v}_1 + \cdots + c_{k-1}(\lambda_{k-1} - \lambda_k)\mathbf{v}_{k-1}.
\end{aligned}$$

and, of course, the right side is **0**. That gives us a linear combination of the first k – 1 vectors which equals **0**, so all their coeffcients are 0:

$$c_1(\lambda_1 - \lambda_k) = \cdots = c_{k-1}(\lambda_{k-1} - \lambda_k) = 0.$$

Since λ_k does not equal any of the other λ_i's, therefore all the c_i's are 0:

$$c_1 = \cdots = c_{k-1} = 0.$$

The original equation now says $c_k\mathbf{v}_k = 0$, and since the eigenvector $\mathbf{v}_k$ is not 0, therefore $c_k = 0$. Thus all k eigenvectors are linearly independent.

Corollary: If a linear operator on an n-dimensional vector space has n distinct eigenvalues, then it's diagonalizable.

Proof: Take an eigenvector for each eigenvalue. By the preceding theorem, they're independent, and since there are n of them, they form a basis of the n-dimensional vector space. The matrix representing the transformation with respect to this basis is diagonal and has the eigenvalues displayed down the diagonal.

Diagonalizability

A linear operator $T : V \to V$ is said to diagonalizable over $\mathbb{F}$ if there exists an ordered basis $\mathcal{B}$ for V such that $[T]_{\mathcal{B}}$ is a diagonal matrix over $\mathbb{F}$. A basis consisting of eigenvectors of T is called an eigenbasis for V. A matrix $A \in M_n(\mathbb{F})$ is called diagonalizable over $\mathbb{F}$ if L_A is a diagonalizable operator over $\mathbb{F}$.

Let $\mathcal{B} = \{v_1, \ldots, v_n\}$ is a basis for V such that $[T]_{\mathcal{B}}$ is the diagonal matrix,

$$\begin{pmatrix} \lambda_1 & & 0 \\ & \ddots & \\ 0 & & \lambda_n \end{pmatrix}$$

then $T(v_i) = \lambda_i v_i$.

Theorem: Let $T : V \to V$ be a linear operator with distinct eigenvalues $\lambda_1, \ldots, \lambda_k$. If $v_1, \ldots, v_k$ are eigenvectors of T corresponding to $\lambda_1, \ldots, \lambda_k$, then $\{v_1, \ldots, v_k\}$ is linearly independent.

Proof: We prove by induction on k.

$k = 1$: $\{v_1\}$ is linearly independent, since $v_1 \neq \mathbf{0}$.

$k \geq 2$: Suppose that the statement is true for $k-1$ distinct eigenvalues, where $k - 1 \geq 1$. Let $v_1, \ldots, v_k$ be eigenvectors corresponding to k-distinct eigenvalues $\lambda_1, \ldots, \lambda_k$. Let,

$$a_1 v_1 + \cdots + a_k v_k = \mathbf{0}.$$

Then,

$$(T - \lambda_k I)(a_1 v_1 + \cdots + a_k v_k) = \mathbf{0};$$

i.e., $a_1(\lambda_1 - \lambda_k)v_1 + a_2(\lambda_2 - \lambda_k)v_2 + \cdots + a_{k-1}(\lambda_{k-1} - \lambda_k)v_{k-1} = \mathbf{0}$.

By the induction hypothesis,

$$a_1(\lambda_1 - \lambda_k)v_1 = a_2(\lambda_2 - \lambda_k)v_2 = \cdots = a_{k-1}(\lambda_{k-1} - \lambda_k)v_{k-1} = \mathbf{0}.$$

Therefore, $a_1 = a_2 = \cdots = a_{k-1} = 0$, since $\lambda_i - \lambda_k \neq 0$ and $v_i \neq \mathbf{0}$ for $1 \leq i \leq k - 1$. Hence, $a_k v_k = \mathbf{0}$. But $v_k \neq \mathbf{0}$, therefore, $a_k = 0$.

Corollary: Let $T : V \to V$ be a linear operator on the vector space V of dimension n over $\mathbb{F}$. If T has n distinct eigenvalues, then T is diagonalizable.

Proof: Let $\lambda_1, \ldots, \lambda_n$ be distinct eigenvalues of T. Let $v_1, \ldots, v_n$ be eigenvectors corresponding to $\lambda_1, \ldots, \lambda_n$ respectively. We know that $\mathcal{B} = \{v_1, \ldots, v_n\}$ is linearly independent and hence $\mathcal{B}$ is a basis for V. Therefore, T is diagonalizable.

Example: Let $T : \mathbb{R}^2 \to \mathbb{R}^2$ be the linear operator,

$$T(x, y) = (x + y, 2y).$$

Then,

$$[T]_{\mathcal{B}} = \begin{pmatrix} 1 & 1 \\ 0 & 2 \end{pmatrix}$$

where $\mathcal{B} = \{(1, 0),(0, 1)\}$. The characteristic polynomial of T is,

$$\chi_T(t) = \det([T]_{\mathcal{B}} - tI_2) = \det\begin{pmatrix} 1-t & 1 \\ 0 & 2-t \end{pmatrix} = (1-t)(2-t).$$

Therefore, T has exactly two distinct eigenvalues. Hence, T is diagonalizable.

Theorem: $T : V \to V$ be a linear operator on the vector space V of dimension n over $\mathbb{F}$. If T is diagonalizable over $\mathbb{F}$, then, the characteristic polynomial splits in $\mathbb{F}$, that is $\chi_T(t) = c(t-\lambda_1)\cdots(t-\lambda_n)$, where $c, \lambda_1, \ldots, \lambda_n$ are in $\mathbb{F}$. In other words, a necessary condition for diagonalizability over $\mathbb{F}$ is that all the eigenvalues of T belong to $\mathbb{F}$.

Proof: If $[T]_{\mathcal{B}} = \begin{pmatrix} \lambda_1 & & 0 \\ & \ddots & \\ 0 & & \lambda_n \end{pmatrix}$, then

$$\begin{aligned} \chi_T(t) &= \det\begin{pmatrix} \lambda_1 - t & & 0 \\ & \ddots & \\ 0 & & \lambda_n - t \end{pmatrix}, \\ &= (-1)^n (t-\lambda_1)(t-\lambda_2)\cdots(t-\lambda_n). \end{aligned}$$

Algebraic and Geometric Multiplicities

Definition: Let λ be an eigenvalue of a linear operator T on the vector space V of dimension n over $\mathbb{F}$. Let $\chi_T(t)$ denote the characteristic polynomial of T:

- The algebraic multiplicity of λ is the largest positive integer m such that $(t - \lambda)$ m is a factor of $\chi_T(t)$.
- The subspace $N_\lambda = \ker(T - \lambda I)$ of V is called the eigenspace of T corresponding to the eigenvalue λ. The dimension of $N_\lambda = \ker(T - \lambda I)$ is called the geometric multiplicity of the eigenvalue λ.

Theorem: Let T be a linear operator on V. Let λ be an eigenvalue of T. Suppose that m and d are the algebraic and geometric multiplicities of λ respectively. Then, $1 \le d \le m$.

Proof: We know that $d = \dim(\ker(T - \lambda I))$. Let $\{v_1, \ldots, v_d\}$ be an ordered basis for the eigenspace $N_\lambda = \ker(T - \lambda I)$. We extend this basis to a basis $\{v_1, \ldots, v_d, v_{d+1}, \ldots, v_n\}$ for V.

We know that $Tv_i = \lambda v_i$ for all $1 \le i \le d$. Therefore,

$$A = [T]_{\mathcal{B}} = \begin{pmatrix} \lambda I_d & B \\ 0 & C \end{pmatrix}.$$

Then,

$$\begin{aligned} \chi_T(t) = \det(A - tI_n) &= \det\begin{pmatrix} (\lambda - t)I_d & B \\ 0 & (c - tI_{n-d}) \end{pmatrix} \\ &= \det((\lambda - t)I_d)\det(c - tI_{n-d}) \\ &= (\lambda - t)^d . g(t); \end{aligned}$$

where $g(t) \in F[t]$. This shows that $(\lambda - t)^d$ is a factor of $\chi_T(t)$ and hence, $d \leq k$.

Example: Let $T : \mathbb{R}_2[x] \to \mathbb{R}_2[x]$ be given by $T(f(x)) = f'(x)$. Then,

$$[T]_{\mathcal{B}} = \begin{pmatrix} 0 & 1 & 0 \\ 0 & 0 & 2 \\ 0 & 0 & 0 \end{pmatrix},$$

where $\mathcal{B} = \{1, x, x^2\}$ is the standard basis for the vector space $\mathbb{R}_2[x]$. We know that the characteristic polynomial of T is $\chi_T(t) = -t^3$ and 0 is the only eigenvalue of T. The algebraic multiplicity of 0 is 3 and the geometric multiplicity of 0 is the number given by,

$$\dim(\ker(T)) = \dim\{f(x) \in \mathbb{R}_2[x] \mid f'(x) = 0\}$$
$$= 1.$$

It follows that T is not diagonalizable since there is no basis for V consisting of eigenvectors of T.

Example: Let T be the operator on $\mathbb{R}^3$ given by,

$$T\begin{pmatrix} a_1 \\ a_2 \\ a_3 \end{pmatrix} = \begin{pmatrix} a_1 & & +a_3 \\ a_1 & +a_2 & +a_3 \\ a_1 & & +a_3 \end{pmatrix}.$$

We determine the eigenspace of T corresponding to each eigenvalue. The matrix representation of T with respect to the standard basis $\mathcal{B}$ is,

$$A = [T]_{\mathcal{B}} = \begin{pmatrix} 1 & 0 & 1 \\ 1 & 1 & 1 \\ 1 & 0 & 1 \end{pmatrix}.$$

The characteristic polynomial $\chi_T(t)$ is $-t(t - 1)(t - 3)$. Therefore, T has three distinct eigenvalues, hence diagonalizable.

The eigenspace N_0: The eigenspace for the eigenvalue 0 is,

$$N_0 = \left\{ \begin{pmatrix} x \\ y \\ z \end{pmatrix} \middle| \; x + y + z = 0; \;\; x + z = 0 \right\}.$$

Therefore, a basis for N_0 is $\left\{ \begin{pmatrix} 1 \\ 0 \\ -1 \end{pmatrix} \right\}$.

The eigenspace N_1: The eigenspace for the eigenvalue 1 is,

$$N_1 = \left\{ \begin{pmatrix} x \\ y \\ z \end{pmatrix} \middle| \; z = 0; \;\; x + z = 0; \;\; x = 0 \right\}.$$

Therefore, a basis for N_1 is $\left\{\begin{pmatrix}0\\1\\0\end{pmatrix}\right\}$.

The eigenspace N_2: The eigenspace for the eigenvalue 2 is,

$$N_2=\left\{\begin{pmatrix}x\\y\\z\end{pmatrix}\middle|\ -x+z=0;\quad x-y+z=0;\quad x-z=0\right\}.$$

Therefore, a basis for N_2 is $\left\{\begin{pmatrix}1\\2\\1\end{pmatrix}\right\}$.

Example: (Friedberg-Insel-Spence)) - Let T be the linear operator on $\mathbb{R}^3$ defined by,

$$T\begin{pmatrix}a_1\\a_2\\a_3\end{pmatrix}=\begin{pmatrix}4a_1 & & +a_3\\2a_1 & +3a_2 & +2a_3\\a_1 & & +4a_3\end{pmatrix}.$$

We determine the eigenspace of T corresponding to each eigenvalue. Let $\mathcal{B}$ be the standard basis for $\mathbb{R}^3$. Then,

$$[T]_{\mathcal{B}}=\begin{pmatrix}4 & 0 & 1\\2 & 3 & 2\\1 & 0 & 4\end{pmatrix}.$$

and hence the characteristic polynomial of T is,

$$\det\left([T]_{\mathcal{B}}-tI\right)=\det\begin{pmatrix}(4-t) & 0 & 1\\2 & (3-t) & 2\\1 & 0 & (4-t)\end{pmatrix}=-(t-5)(t-3)^2.$$

So the eigenvalues of T are $\lambda_1=5$ and $\lambda_2=3$ with multiplicities 1 and 2, respectively.

The eigenspace corresponding to $\lambda_1=1$ is,

$$E_{\lambda_1}=N(T-\lambda_1 I)=\left\{\begin{pmatrix}x_1\\x_2\\x_3\end{pmatrix}\in\mathbb{R}^3\ \middle|\ \begin{pmatrix}-1 & 0 & 1\\2 & -2 & 2\\1 & 0 & -1\end{pmatrix}\begin{pmatrix}x_1\\x_2\\x_3\end{pmatrix}=\begin{pmatrix}0\\0\\0\end{pmatrix}\right\}.$$

E_{λ_1} is the solution space for the system of linear equations,

$$-x_1+x_3=0$$
$$2x_1-2x_2+2x_3=0$$
$$x_1-x_3=0.$$

It is easily seen that $\left\{\begin{pmatrix}1\\2\\1\end{pmatrix}\right\}$

is a basis for E_{λ_1}. Hence dim $E_{\lambda_1} = 1$.

Similarly, $E_{\lambda_2} = N(T - \lambda_2 I)$ is the solution space of the system,

$$\begin{aligned} x_1 + x_3 &= 0 \\ 2x_1 + 2x_3 &= 0 \\ x_1 + x_3 &= 0. \end{aligned}$$

Since, the unknown x_2 does not appear in the system, we assign it a parametric value, say, $x_2 = s$, and solve the system for x_1 and x_3, introducing another parameter t. The result is the general solution to the system,

$$\begin{pmatrix}x_1\\x_2\\x_3\end{pmatrix} = s\begin{pmatrix}0\\1\\0\end{pmatrix} + t\begin{pmatrix}-1\\0\\1\end{pmatrix}$$

for $s, t \in \mathbb{R}$ it follows that,

$$\left\{\begin{pmatrix}0\\1\\0\end{pmatrix}\begin{pmatrix}-1\\0\\1\end{pmatrix}\right\}$$

is a basis for E_{λ_2} and dim $E_{\lambda_2} = 2$.

In this case the multiplicity of each eigenvalue λ_i is equal to the dimension of the corresponding eigenspace E_{λ_i}. Observe that the union of the two bases just derived, namely,

$$\left\{\begin{pmatrix}1\\2\\1\end{pmatrix}, \begin{pmatrix}0\\1\\0\end{pmatrix}, \begin{pmatrix}-1\\0\\1\end{pmatrix}\right\}$$

is linearly independent and hence is a basis for $\mathbb{R}^3$ consisting of eigenvectors of T. Consequently, T is diagonalizable.

References

- Vector-Space: mathworld.wolfram.com, Retrieved 11, May 2020
- Linear-combinations, matrix-algebra: statlect.com, Retrieved 27, July 2020
- Linear-Transformation: mathworld.wolfram.com, Retrieved 19, March 2020
- Introduction-to-matrices, boundless-algebra: courses.lumenlearning.com, Retrieved 17, August 2020
- linear-algebra, the-rank-of-a-matrix, real-euclidean-vector-spaces, algebra: cliffsnotes.com, Retrieved 25, February 2020

4

Algebraic Equations

Algebraic equations are the equations in which two expressions are set equal to each other. Polynomials are the expressions that contain variables and coefficients and involve the operations of multiplication, addition, subtraction and positive integer exponentiation of variables. This chapter closely examines the key concepts of algebraic equations to provide an extensive understanding of the subject.

In mathematics, an algebraic equation or polynomial equation is an equation of the form

$$P = Q$$

where P and Q are polynomials with coefficients in some field, often the field of the rational numbers. For most authors, an algebraic equation is *univariate*, which means that it involves only one variable. On the other hand, a polynomial equation may involve several variables, in which case it is called *multivariate* and the term *polynomial equation* is usually preferred to *algebraic equation*.

For example,

$$x^5 - 3x + 1 = 0.$$

is an algebraic equation with integer coefficients and

$$y^4 + \frac{xy}{2} = \frac{x^3}{3} - xy^2 + y^2 - \frac{1}{7}.$$

is a multivariate polynomial equation over the rationals.

Some but not all polynomial equations with rational coefficients have a solution that is an algebraic expression that can be found using a finite number of operations that involve only those same types of coefficients (that is, can be solved algebraically). This can be done for all such equations of degree one, two, three, or four; but for degree five or more it can only be done for some equations, not for all. A large amount of research has been devoted to compute efficiently accurate approximations of the real or complex solutions of a univariate algebraic equation and of the common solutions of several multivariate polynomial equations.

History

The study of algebraic equations is probably as old as mathematics: the Babylonian mathematicians, as early as 2000 BC could solve some kinds of quadratic equations (displayed on Old Babylonian clay tablets).

Univariate algebraic equations over the rationals (i.e., with rational coefficients) have a very long history. Ancient mathematicians wanted the solutions in the form of radical expressions, like $x = \frac{1+\sqrt{5}}{2}$ for the positive solution of $x^2 - x - 1 = 0$. The ancient Egyptians knew how to solve equations of degree 2 in this manner. The Indian mathematician Brahmagupta (597–668 AD) explicitly described the quadratic formula in his treatise Brāhmasphuṭasiddhānta published in 628 AD, but written in words instead of symbols. In the 9th century Muhammad ibn Musa al-Khwarizmi and other Islamic mathematicians derived the quadratic formula, the general solution of equations of degree 2, and recognized the importance of the discriminant. During the Renaissance in 1545, Gerolamo Cardano published the solution of Scipione del Ferro and Niccolò Fontana Tartaglia to equations of degree 3 and that of Lodovico Ferrari for equations of degree 4. Finally Niels Henrik Abel proved, in 1824, that equations of degree 5 and higher do not have general solutions using radicals. Galois theory, named after Évariste Galois, showed that some equations of at least degree 5 do not even have an idiosyncratic solution in radicals, and gave criteria for deciding if an equation is in fact solvable using radicals.

Areas of Study

The algebraic equations are the basis of a number of areas of modern mathematics: Algebraic number theory is the study of (univariate) algebraic equations over the rationals (that is, with rational coefficients). Galois theory has been introduced by Évariste Galois for getting criteria deciding if an algebraic equation may be solved in terms of radicals. In field theory, an algebraic extension is an extension such that every element is a root of an algebraic equation over the base field. Transcendental number theory is the study of the real numbers which are not solutions to an algebraic equation over the rationals. A Diophantine equation is a (usually multivariate) polynomial equation with integer coefficients for which one is interested in the integer solutions. Algebraic geometry is the study of the solutions in an algebraically closed field of multivariate polynomial equations.

Two equations are equivalent if they have the same set of solutions. In particular the equation $P = Q$ is equivalent with $P - Q = 0$. It follows that the study of algebraic equations is equivalent to the study of polynomials.

A polynomial equation over the rationals can always be converted to an equivalent one in which the coefficients are integers. For example, multiplying through by 42 = 2·3·7 and grouping its terms in the first member, the previously mentioned polynomial equation $y^4 + \frac{xy}{2} = \frac{x^3}{3} - xy^2 + y^2 - \frac{1}{7}$ becomes,

$$42y^4 + 21xy - 14x^3 + 42xy^2 - 42y^2 + 6 = 0.$$

Because sine, exponentiation, and $1/T$ are not polynomial functions,

$$e^T x^2 + \frac{1}{T}xy + \sin(T)z - 2 = 0.$$

is *not* a polynomial equation in the four variables *x*, *y*, *z*, and *T* over the rational numbers. However, it is a polynomial equation in the three variables *x*, *y*, and *z* over the field of the elementary functions in the variable *T*.

Solutions

As for any equation, the *solutions* of an equation are the values of the variables for which the equation is true. For univariate algebraic equations these are also called roots, even if, properly speaking, one should say *the solutions of the algebraic equation P=0 are the* ***roots*** *of the polynomial P*. When solving an equation, it is important to specify in which set the solutions are allowed. For example, for an equation over the rationals one may look for solutions in which all the variables are integers. In this case the equation is a Diophantine equation. One may also be interested only in the real solutions. However, for univariate algebraic equations, the number of solutions is finite, and all solutions are contained in any algebraically closed field containing the coefficients—for example, the field of complex numbers in the case of equations over the rationals. It follows that without precision "root" and "solution" usually mean "solution in an algebraically closed field".

Linear Equation

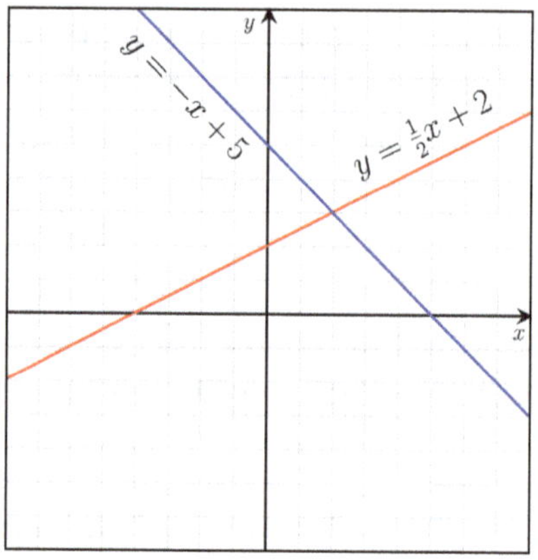

Graph sample of linear equations.

A linear equation is an algebraic equation in which each term is either a constant or the product of a constant and (the first power of) a single variable. A simple example of a linear equation with only one variable, *x*, may be written in the form: $ax + b = 0$, where *a* and *b* are constants and $a \neq 0$. The constants may be numbers, parameters, or even non-linear functions of parameters, and the distinction between variables and parameters may depend on the problem.

Linear equations can have one or more variables. An example of a linear equation with three variables, *x*, *y*, and *z*, is given by: $ax + by + cz + d = 0$, where *a*, *b*, *c*, and *d* are constants and *a*, *b*, and *c* are non-zero. Linear equations occur frequently in most subareas of mathematics and especially in applied mathematics. While they arise quite naturally when modeling many phenomena, they are particularly useful since many non-linear equations may be reduced to linear equations by assuming that quantities of interest vary to only a small extent from some "background" state. An equation is linear if the sum of the exponents of the variables of each term is one.

Equations with exponents greater than one are non-linear. An example of a non-linear equation of two variables is $axy + b = 0$, where a and b are constants and $a \neq 0$. It has two variables, x and y, and is non-linear because the sum of the exponents of the variables in the first term, axy, is two.

This article considers the case of a single equation for which one searches the real solutions. All its content applies for complex solutions and, more generally for linear equations with coefficients and solutions in any field.

One Variable

A linear equation in one unknown x may always be rewritten

$$ax = b.$$

If $a \neq 0$, there is a unique solution

$$x = \frac{b}{a}.$$

If $a = 0$, then, when $b = 0$ every number is a solution of the equation, but if $b \neq 0$ there are no solutions (and the equation is said to be inconsistent.)

Two Variables

A common form of a linear equation in the two variables x and y is

$$y = mx + b,$$

where m and b designate constants (parameters). The origin of the name "linear" comes from the fact that the set of solutions of such an equation forms a straight line in the plane. In this particular equation, the constant m determines the slope or gradient of that line, and the constant term b determines the point at which the line crosses the y-axis, otherwise known as the y-intercept.

Since terms of linear equations cannot contain products of distinct or equal variables, nor any power (other than 1) or other function of a variable, equations involving terms such as xy, x^2, $y^{1/3}$, and $\sin(x)$ are *nonlinear*.

Forms for Two-dimensional Linear Equations

Linear equations can be rewritten using the laws of elementary algebra into several different forms. These equations are often referred to as the "equations of the straight line." In what follows, x, y, t, and θ are variables; other letters represent constants (fixed numbers).

General (or Standard) Form

In the general (or standard) form the linear equation is written as:

$$Ax + By = C,$$

where A and B are not both equal to zero. The equation is usually written so that $A \geq 0$, by convention. The graph of the equation is a straight line, and every straight line can be represented by an equation in the above form. If A is nonzero, then the x-intercept, that is, the x-coordinate of the point where the graph crosses the x-axis (where, y is zero), is C/A. If B is nonzero, then the y-intercept, that is the y-coordinate of the point where the graph crosses the y-axis (where x is zero), is C/B, and the slope of the line is $-A/B$. The general form is sometimes written as:

$$ax + by + c = 0,$$

where a and b are not both equal to zero. The two versions can be converted from one to the other by moving the constant term to the other side of the equal sign.

Slope–intercept Form

$$y = mx + b,$$

where m is the slope of the line and b is the y intercept, which is the y coordinate of the location where the line crosses the y axis. This can be seen by letting $x = 0$, which immediately gives $y = b$. It may be helpful to think about this in terms of $y = b + mx$; where the line passes through the point $(0, b)$ and extends to the left and right at a slope of m. Vertical lines, having undefined slope, cannot be represented by this form.

Point–slope Form

$$y - y_1 = m(x - x_1),$$

where m is the slope of the line and (x_1,y_1) is any point on the line.

The point-slope form expresses the fact that the difference in the y coordinate between two points on a line (that is, $y - y_1$) is proportional to the difference in the x coordinate (that is, $x - x_1$). The proportionality constant is m (the slope of the line).

Two-point Form

$$y - y_1 = \frac{y_2 - y_1}{x_2 - x_1}(x - x_1),$$

where (x_1, y_1) and (x_2, y_2) are two points on the line with $x_2 \neq x_1$. This is equivalent to the point-slope form above, where the slope is explicitly given as $(y_2 - y_1)/(x_2 - x_1)$.

Multiplying both sides of this equation by $(x_2 - x_1)$ yields a form of the line generally referred to as the symmetric form:

$$(x_2 - x_1)(y - y_1) = (y_2 - y_1)(x - x_1).$$

Expanding the products and regrouping the terms leads to the general form:

$$x(y_2 - y_1) - y(x_2 - x_1) = x_1 y_2 - x_2 y_1.$$

Using a determinant, one gets a determinant form, easy to remember:

$$\begin{vmatrix} x & y & 1 \\ x_1 & y_1 & 1 \\ x_2 & y_2 & 1 \end{vmatrix} = 0.$$

Intercept Form

$$\frac{x}{a} + \frac{y}{b} = 1,$$

where a and b must be nonzero. The graph of the equation has x-intercept a and y-intercept b. The intercept form is in standard form with $A/C = 1/a$ and $B/C = 1/b$. Lines that pass through the origin or which are horizontal or vertical violate the nonzero condition on a or b and cannot be represented in this form.

Matrix Form

Using the order of the standard form

$$Ax + By = C,$$

one can rewrite the equation in matrix form:

$$\begin{pmatrix} A & B \end{pmatrix} \begin{pmatrix} x \\ y \end{pmatrix} = \begin{pmatrix} C \end{pmatrix}.$$

Further, this representation extends to systems of linear equations.

$$A_1 x + B_1 y = C_1,$$

$$A_2 x + B_2 y = C_2,$$

becomes:

$$\begin{pmatrix} A_1 & B_1 \\ A_2 & B_2 \end{pmatrix} \begin{pmatrix} x \\ y \end{pmatrix} = \begin{pmatrix} C_1 \\ C_2 \end{pmatrix}.$$

Since this extends easily to higher dimensions, it is a common representation in linear algebra, and in computer programming. There are named methods for solving system of linear equations, like Gauss-Jordan which can be expressed as matrix elementary row operations.

Parametric Form

$$x = Tt + U$$

and

$$y = Vt + W.$$

Two simultaneous equations in terms of a variable parameter *t*, with slope $m = V / T$, *x*-intercept $(VU - WT) / V$ and *y*-intercept $(WT - VU) / T$. This can also be related to the two-point form, where $T = p - h$, $U = h$, $V = q - k$, and $W = k$:

$$x = (p - h)t + h$$

and

$$y = (q - k)t + k.$$

In this case *t* varies from 0 at point (*h*,*k*) to 1 at point (*p*,*q*), with values of *t* between 0 and 1 providing interpolation and other values of *t* providing extrapolation.

2D Vector Determinant Form

The equation of a line can also be written as the determinant of two vectors. If P_1 and P_2 are unique points on the line, then P will also be a point on the line if the following is true:

$$\det(\overrightarrow{P_1P}, \overrightarrow{P_1P_2}) = 0.$$

One way to understand this formula is to use the fact that the determinant of two vectors on the plane will give the area of the parallelogram they form. Therefore, if the determinant equals zero then the parallelogram has no area, and that will happen when two vectors are on the same line.

To expand on this we can say that $P_1 = (x_1, y_1)$, $P_2 = (x_2, y_2)$ and $P = (x, y)$. Thus $\overrightarrow{P_1P} = (x - x_1, y - y_1)$ and $\overrightarrow{P_1P_2} = (x_2 - x_1, y_2 - y_1)$, then the above equation becomes:

$$\det\begin{pmatrix} x - x_1 & y - y_1 \\ x_2 - x_1 & y_2 - y_1 \end{pmatrix} = 0.$$

Thus,

$$(x - x_1)(y_2 - y_1) - (y - y_1)(x_2 - x_1) = 0.$$

Ergo,

$$(x - x_1)(y_2 - y_1) = (y - y_1)(x_2 - x_1).$$

Then dividing both side by $(x_2 - x_1)$ would result in the "Two-point form" shown above, but leaving it here allows the equation to still be valid when $x_1 = x_2$.

Special Cases

$y = b$

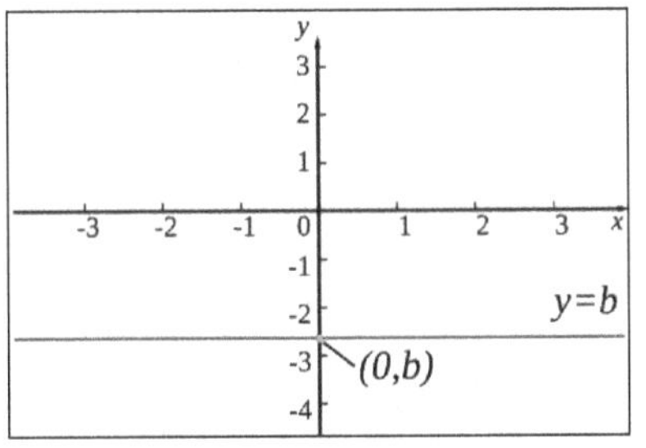

Horizontal Line $y = b$.

This is a special case of the standard form where $A = 0$ and $B = 1$, or of the slope-intercept form where the slope $m = 0$. The graph is a horizontal line with y-intercept equal to b. There is no x-intercept, unless $b = 0$, in which case the graph of the line is the x-axis, and so every real number is an x-intercept.

$x = a$

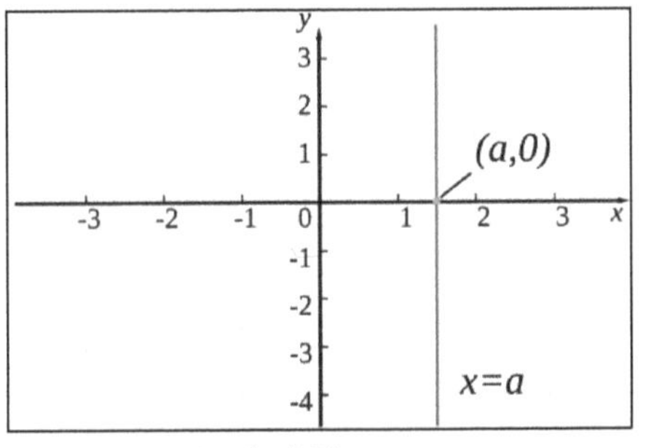

Vertical Line $x = a$.

This is a special case of the standard form where $A = 1$ and $B = 0$. The graph is a vertical line with x-intercept equal to a. The slope is undefined. There is no y-intercept, unless $a = 0$, in which case the graph of the line is the y-axis, and so every real number is a y-intercept. This is the only type of straight line which is not the graph of a function (it obviously fails the vertical line test).

Connection with Linear Functions

A linear equation, written in the form $y = f(x)$ whose graph crosses the origin $(x,y) = (0,0)$, that is, whose y-intercept is 0, has the following properties:

$$f(x_1 + x_2) = f(x_1) + f(x_2)$$

and

$$f(ax) = af(x),$$

where a is any scalar. A function which satisfies these properties is called a *linear function* (or *linear operator*, or more generally a *linear map*). However, linear equations that have non-zero y-intercepts, when written in this manner, produce functions which will have neither property above and hence are not linear functions in this sense. They are known as affine functions.

Examples

An everyday example of the use of different forms of linear equations is computation of tax with tax brackets. This is commonly done using either point–slope form or slope–intercept form.

More than Two Variables

A linear equation can involve more than two variables. Every linear equation in n unknowns may be rewritten.

$$a_1x_1 + a_2x_2 + \cdots + a_nx_n = b,$$

where, $a_1, a_2, \ldots, a_n$ represent numbers, called the *coefficients*, $x_1, x_2, \ldots, x_n$ are the unknowns, and b is called the *constant term*. When dealing with three or fewer variables, it is common to use x, y and z instead of x_1, x_2 and x_3.

If all the coefficients are zero, then either $b \neq 0$ and the equation does not have any solution, or $b = 0$ and every set of values for the unknowns is a solution.

If at least one coefficient is nonzero, a permutation of the subscripts allows to suppose $a_1 \neq 0$, and rewrite the equation.

$$x_1 = \frac{b}{a_1} - \frac{a_2}{a_1}x_2 - \cdots - \frac{a_n}{a_1}x_n.$$

In other words, if $a_i \neq 0$, one may choose arbitrary values for all the unknowns except x_i, and express x_i in term of these values.

If $n = 3$ the set of the solutions is a plane in a three-dimensional space. More generally, the set of the solutions is an $(n - 1)$-dimensional hyperplane in a n-dimensional Euclidean space (or affine space if the coefficients are complex numbers or belong to any field).

Polynomial

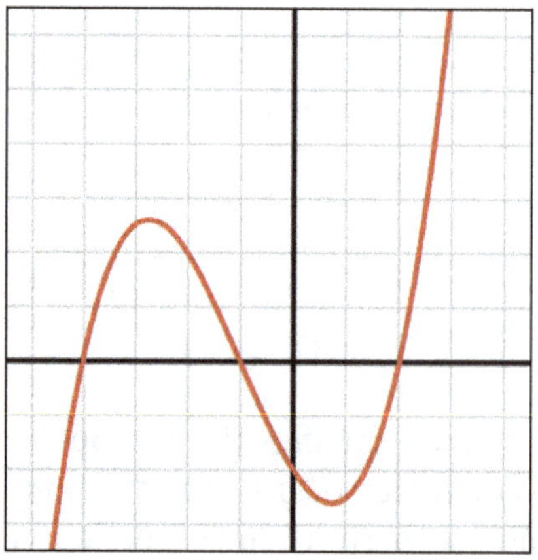

The graph of a polynomial function of degree 3.

In mathematics, a polynomial is an expression consisting of variables (or indeterminates) and coefficients, that involves only the operations of addition, subtraction, multiplication, and non-negative integer exponents. An example of a polynomial of a single indeterminate x is $x^2 - 4x + 7$. An example in three variables is $x^3 + 2xyz^2 - yz + 1$.

Polynomials appear in a wide variety of areas of mathematics and science. For example, they are used to form polynomial equations, which encode a wide range of problems, from elementary word problems to complicated problems in the sciences; they are used to define polynomial functions, which appear in settings ranging from basic chemistry and physics to economics and social science; they are used in calculus and numerical analysis to approximate other functions. In advanced mathematics, polynomials are used to construct polynomial rings and algebraic varieties, central concepts in algebra and algebraic geometry.

Etymology

The word *polynomial* joins two diverse roots: the Greek *poly*, meaning "many," and the Latin *nomen*, or name. It was derived from the term *binomial* by replacing the Latin root *bi-* with the Greek *poly-*. The word *polynomial* was first used in the 17th century.

Notation and Terminology

The x occurring in a polynomial is commonly called either a *variable* or an *indeterminate*. When the polynomial is considered as an expression, x is a fixed symbol which does not have any value (its value is "indeterminate"). It is thus more correct to call it an "indeterminate". However, when one considers the function defined by the polynomial, then x represents the argument of the function, and is therefore called a "variable". Many authors use these two words interchangeably.

It is a common convention to use uppercase letters for the indeterminates and the corresponding lowercase letters for the variables (arguments) of the associated function.

It may be confusing that a polynomial P in the indeterminate x may appear in the formulas either as P or as $P(x)$.

Normally, the name of the polynomial is P, not $P(x)$. However, if a denotes a number, a variable, another polynomial, or, more generally any expression, then $P(a)$ denotes, by convention, the result of substituting x by a in P. Thus, the polynomial P defines the function.

$$a \mapsto P(a),$$

which is the polynomial function associated to P.

Frequently, when using this function, one supposes that a is a number. However one may use it over any domain where addition and multiplication are defined (any ring). In particular, when a is the indeterminate x, then the image of x by this function is the polynomial P itself (substituting x to x does not change anything). In other words,

$$P(x) = P.$$

This equality allows writing "let $P(x)$ be a polynomial" as a shorthand for "let P be a polynomial in the indeterminate x". On the other hand, when it is not necessary to emphasize the name of the indeterminate, many formulas are much simpler and easier to read if the name(s) of the indeterminate(s) do not appear at each occurrence of the polynomial.

Definition

A polynomial is an expression that can be built from constants and symbols called indeterminates or variables by means of addition, multiplication and exponentiation to a non-negative power. Two such expressions that may be transformed, one to the other, by applying the usual properties of commutativity, associativity and distributivity of addition and multiplication are considered as defining the same polynomial.

A polynomial in a single indeterminate x can always be written (or rewritten) in the form

$$a_n x^n + a_{n-1} x^{n-1} + \cdots + a_2 x^2 + a_1 x + a_0,$$

where $a_0, \ldots, a_n$ are constants and x is the indeterminate. The word "indeterminate" means that x represents no particular value, although any value may be substituted for it. The mapping that associates the result of this substitution to the substituted value is a function, called a *polynomial function*.

This can be expressed more concisely by using summation notation:

$$\sum_{i=0}^{n} a_i x^i$$

That is, a polynomial can either be zero or can be written as the sum of a finite number of non-zero terms. Each term consists of the product of a number—called the coefficient of the term—and a finite number of indeterminates, raised to nonnegative integer powers. The exponent on an indeterminate in a term is called the degree of that indeterminate in that term; the degree of the term is the sum of the degrees of the indeterminates in that term, and the degree of a polynomial is the largest degree of any one term with nonzero coefficient. Because $x = x^1$, the degree of an indeterminate without a written exponent is one.

A term and a polynomial with no indeterminates are called, respectively, a constant term and a constant polynomial. The degree of a constant term and of a nonzero constant polynomial is 0. The degree of the zero polynomial, 0, (which has no terms at all) is generally treated as not defined.

For example:

$$-5x^2y$$

is a term. The coefficient is −5, the indeterminates are x and y, the degree of x is two, while the degree of y is one. The degree of the entire term is the sum of the degrees of each indeterminate in it, so in this example the degree is $2 + 1 = 3$.

Forming a sum of several terms produces a polynomial. For example, the following is a polynomial:

$$\underbrace{-3x^2}_{\substack{\text{term}\\1}}\underbrace{-5x}_{\substack{\text{term}\\2}}\underbrace{+4}_{\substack{\text{term}\\3}}.$$

It consists of three terms: the first is degree two, the second is degree one, and the third is degree zero.

Polynomials of small degree have been given specific names. A polynomial of degree zero is a *constant polynomial* or simply a *constant.* Polynomials of degree one, two or three are respectively *linear polynomials, quadratic polynomials* and *cubic polynomials.* For higher degrees the specific names are not commonly used, although *quartic polynomial* (for degree four) and *quintic polynomial* (for degree five) are sometimes used. The names for the degrees may be applied to the polynomial or to its terms. For example, in $x^2 + 2x + 1$ the term $2x$ is a linear term in a quadratic polynomial.

The polynomial 0, which may be considered to have no terms at all, is called the zero polynomial. Unlike other constant polynomials, its degree is not zero. Rather the degree of the zero polynomial is either left explicitly undefined, or defined as negative (either -1 or $-\infty$). These conventions are useful when defining Euclidean division of polynomials. The zero polynomial is also unique in that it is the only polynomial having an infinite number of roots. The graph of the zero polynomial, $f(x) = 0$, is the X-axis.

In the case of polynomials in more than one indeterminate, a polynomial is called *homogeneous* of degree n if *all* its non-zero terms have degree n. The zero polynomial is homogeneous, and, as homogeneous polynomial, its degree is undefined. For example, $x^3y^2 + 7x^2y^3 - 3x^5$ is homogeneous of degree 5.

The commutative law of addition can be used to rearrange terms into any preferred order. In polynomials with one indeterminate, the terms are usually ordered according to degree, either in "descending powers of x", with the term of largest degree first, or in "ascending powers of x". The polynomial in the example above is written in descending powers of x. The first term has coefficient 3, indeterminate x, and exponent 2. In the second term, the coefficient is -5. The third term is a constant. Because the *degree* of a non-zero polynomial is the largest degree of any one term, this polynomial has degree two.

Two terms with the same indeterminates raised to the same powers are called "similar terms" or "like terms", and they can be combined, using the distributive law, into a single term whose coefficient is the sum of the coefficients of the terms that were combined. It may happen that this makes the coefficient 0. Polynomials can be classified by the number of terms with nonzero coefficients, so that a one-term polynomial is called a monomial, a two-term polynomial is called a binomial, and a three-term polynomial is called a *trinomial.* The term "quadrinomial" is occasionally used for a four-term polynomial.

A polynomial in one indeterminate is called a *univariate polynomial,* a polynomial in more than one indeterminate is called a multivariate polynomial. A polynomial with two indeterminates is called a bivariate polynomial. These notions refer more to the kind of polynomials one is generally working with than to individual polynomials; for instance when working with univariate polynomials one does not exclude constant polynomials (which may result, for instance, from the subtraction of non-constant polynomials), although strictly speaking constant polynomials do not contain any indeterminates at all. It is possible to further classify multivariate polynomials as *bivariate*, *trivariate*, and so on, according to the maximum number of indeterminates allowed. Again, so that the set of objects under consideration be closed under subtraction, a study of trivariate polynomials usually allows bivariate polynomials, and so on. It is common, also, to say simply "polynomials in x, y, and z", listing the indeterminates allowed.

The *evaluation of a polynomial* consists of substituting a numerical value to each indeterminate and carrying out the indicated multiplications and additions. For polynomials in one indeterminate, the evaluation is usually more efficient (lower number of arithmetic operations to perform) using Horner's method:

$$(((\cdots((a_n x + a_{n-1})x + a_{n-2})x + \cdots + a_3)x + a_2)x + a_1)x + a_0.$$

Arithmetic

Polynomials can be added using the associative law of addition (grouping all their terms together into a single sum), possibly followed by reordering, and combining of like terms. For example, if

$$P = 3x^2 - 2x + 5xy - 2$$
$$Q = -3x^2 + 3x + 4y^2 + 8$$

then

$$P + Q = 3x^2 - 2x + 5xy - 2 - 3x^2 + 3x + 4y^2 + 8$$

which can be simplified to

$$P + Q = x + 5xy + 4y^2 + 6$$

To work out the product of two polynomials into a sum of terms, the distributive law is repeatedly applied, which results in each term of one polynomial being multiplied by every term of the other. For example, if;

$$P = 2x + 3y + 5$$
$$Q = 2x + 5y + xy + 1$$

then

$$\begin{array}{rcccccccc} PQ = & (2x\cdot 2x) & + & (2x\cdot 5y) & + & (2x\cdot xy) & + & (2x\cdot 1) \\ + & (3y\cdot 2x) & + & (3y\cdot 5y) & + & (3y\cdot xy) & + & (3y\cdot 1) \\ + & (5\cdot 2x) & + & (5\cdot 5y) & + & (5\cdot xy) & + & (5\cdot 1) \end{array}$$

which can be simplified to,

$$PQ = 4x^2 + 21xy + 2x^2y + 12x + 15y^2 + 3xy^2 + 28y + 5$$

Polynomial evaluation can be used to compute the remainder of polynomial division by a polynomial of degree one, because the remainder of the division of $f(x)$ by $(x - a)$ is $f(a)$. This is more efficient than the usual algorithm of division when the quotient is not needed.

A sum of polynomials is a polynomial.

A product of polynomials is a polynomial.

A composition of two polynomials is a polynomial, which is obtained by substituting a variable of the first polynomial by the second polynomial.

- The derivative of the polynomial $a_n x^n + a_{n-1} x^{n-1} + \ldots + a_2 x^2 + a_1 x + a_0$ is the polynomial $n a_n x^{n-1} + (n-1) a_{n-1} x^{n-2} + \ldots + 2 a_2 x + a_1$. If the set of the coefficients does not contain the integers (for example if the coefficients are integers modulo some prime number p), then $k a_k$ should be interpreted as the sum of a_k with itself, k times. For example, over the integers modulo p, the derivative of the polynomial $x^p + 1$ is the polynomial 0.
- A primitive integral or antiderivative of the polynomial $a_n x^n + a_{n-1} x^{n-1} + \ldots + a_2 x^2 + a_1 x + a_0$ is the polynomial $a_n x^{n+1}/(n+1) + a_{n-1} x^n/n + \ldots + a_2 x^3/3 + a_1 x^2/2 + a_0 x + c$, where c is an arbitrary constant. For instance, the antiderivatives of $x^2 + 1$ have the form $1/3x^3 + x + c$.

As for the integers, two kinds of divisions are considered for the polynomials. The *Euclidean division of polynomials* that generalizes the Euclidean division of the integers. It results in two polynomials, a *quotient* and a *remainder* that are characterized by the following property of the polynomials: given two polynomials a and b such that $b \neq 0$, there exists a unique pair of polynomials, q, the quotient, and r, the remainder, such that $a = b\,q + r$ and degree(r) < degree(b) (here the polynomial zero is supposed to have a negative degree). By hand as well as with a computer, this division can be computed by the polynomial long division algorithm.

All polynomials with coefficients in a unique factorization domain (for example, the integers or a field) also have a factored form in which the polynomial is written as a product of irreducible polynomials and a constant. This factored form is unique up to the order of the factors and their multiplication by an invertible constant. In the case of the field of complex numbers, the irreducible factors are linear. Over the real numbers, they have the degree either one or two. Over the integers and the rational numbers the irreducible factors may have any degree. For example, the factored form of,

$$5x^3 - 5$$

is

$$5(x-1)\left(x^2 + x + 1\right)$$

over the integers and the reals and

$$5(x-1)\left(x + \frac{1 + i\sqrt{3}}{2}\right)\left(x + \frac{1 - i\sqrt{3}}{2}\right)$$

over the complex numbers.

The computation of the factored form, called *factorization* is, in general, too difficult to be done by hand-written computation. However, efficient polynomial factorization algorithms are available in most computer algebra systems.

A formal quotient of polynomials, that is, an algebraic fraction wherein the numerator and denominator are polynomials, is called a "rational expression" or "rational fraction" and is not, in general, a polynomial. Division of a polynomial by a number, however, yields another polynomial. For example, $x^3/12$ is considered a valid term in a polynomial (and a polynomial by itself) because it is equivalent to $(1/12)x^3$ and $1/12$ is just a constant. When this expression is used as a term, its coefficient is therefore $1/12$. For similar reasons, if complex coefficients are allowed, one may have a single term like $(2 + 3i)\,x^3$; even though it looks like it should be expanded to two terms, the complex number $2 + 3i$ is one complex number, and is the coefficient of that term. The expression $1/(x^2 + 1)$ is not a polynomial because it includes division by a non-constant polynomial. The expression $(5 + y)^x$ is not a polynomial, because it contains an indeterminate used as exponent.

Because subtraction can be replaced by addition of the opposite quantity, and because positive integer exponents can be replaced by repeated multiplication, all polynomials can be constructed from constants and indeterminates using only addition and multiplication.

Polynomial Functions

A *polynomial function* is a function that can be defined by evaluating a polynomial. A function f of one argument is thus a polynomial function if it satisfies.

$$f(x) = a_n x^n + a_{n-1} x^{n-1} + \cdots + a_2 x^2 + a_1 x + a_0$$

for all arguments x, where n is a non-negative integer and $a_0, a_1, a_2, \ldots, a_n$ are constant coefficients.

For example, the function f, taking real numbers to real numbers, defined by:

$$f(x) = x^3 - x$$

is a polynomial function of one variable. Polynomial functions of multiple variables are similarly defined, using polynomials in multiple indeterminates, as in:

$$f(x, y) = 2x^3 + 4x^2 y + xy^5 + y^2 - 7.$$

An example is also the function $f(x) = \cos(2\arccos(x))$ which, although it does not look like a polynomial, is a polynomial function on $[-1,1]$ because for every x from $f(x) = 2x^2 - 1$ it is true that $f(x) = 2x^2 - 1$.

Polynomial functions are a class of functions having many important properties. They are all continuous, smooth, entire, computable, etc.

Graphs

- 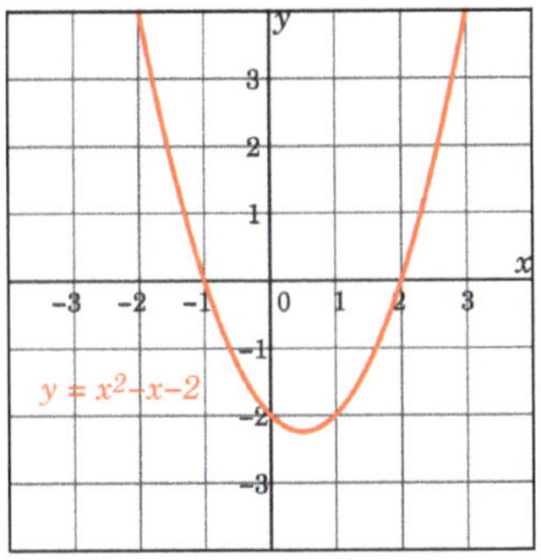

  Polynomial of degree 2:
  $f(x) = x^2 - x - 2$
  $= (x + 1)(x - 2)$.

- 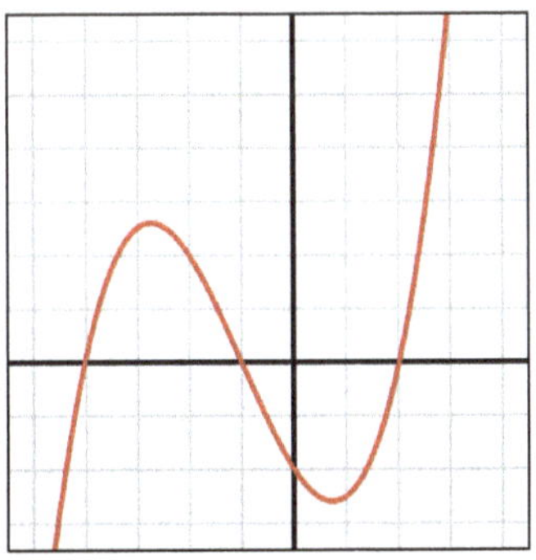

  Polynomial of degree 3:
  $f(x) = x^3/4 + 3x^2/4 - 3x/2 - 2$
  $= 1/4\ (x + 4)(x + 1)(x - 2)$.

- 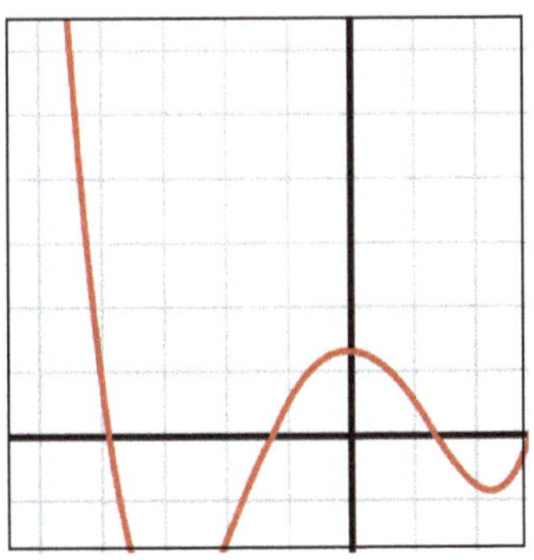

  Polynomial of degree 4:
  $f(x) = 1/14\ (x + 4)(x + 1)(x - 1)(x - 3)$
  $+ 0.5$.

- 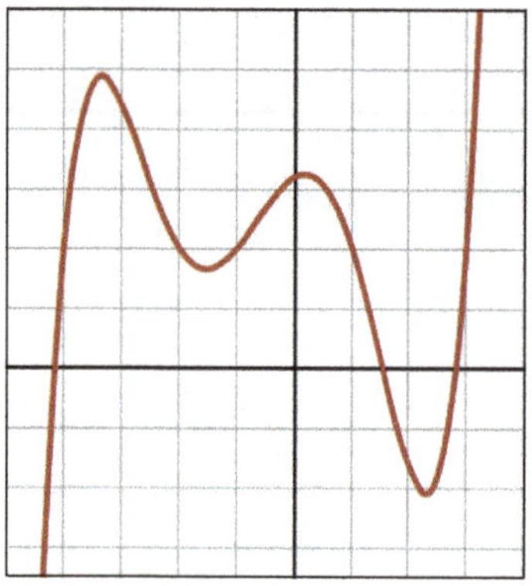

Polynomial of degree 5:
$f(x) = 1/20\ (x + 4)(x + 2)(x + 1)(x - 1)$
$(x - 3) + 2$.

- 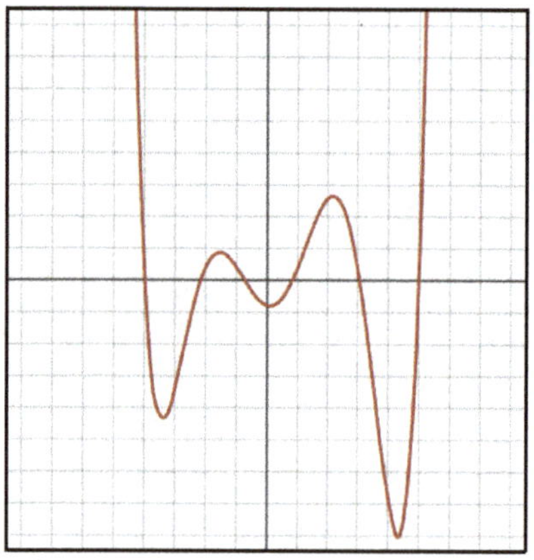

Polynomial of degree 6:
$f(x) = 1/100\ (x^6 - 2x^5 - 26x^4 + 28x^3$
$+ 145x^2 - 26x - 80)$.

- 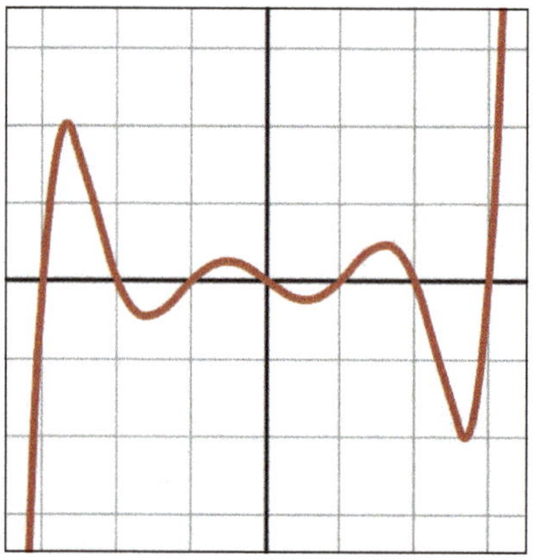

Polynomial of degree 7:
$f(x) = (x - 3)(x - 2)(x - 1)(x)(x + 1)(x + 2)$
$(x + 3)$.

A polynomial function in one real variable can be represented by a graph.

- The graph of the zero polynomial

  $f(x) = 0$

  is the x-axis.
- The graph of a degree 0 polynomial

  $f(x) = a_0$, where $a_0 \neq 0$,

  is a horizontal line with y-intercept a_0
- The graph of a degree 1 polynomial (or linear function)

  $f(x) = a_0 + a_1 x$, where $a_1 \neq 0$,

  is an oblique line with y-intercept a_0 and slope a_1.
- The graph of a degree 2 polynomial

  $f(x) = a_0 + a_1 x + a_2 x^2$, where $a_2 \neq 0$

  is a parabola.
- The graph of a degree 3 polynomial

  $f(x) = a_0 + a_1 x + a_2 x^2 + a_3 x^3$, where $a_3 \neq 0$

  is a cubic curve.
- The graph of any polynomial with degree 2 or greater

  $f(x) = a_0 + a_1 x + a_2 x^2 + \ldots + a_n x^n$, where $a_n \neq 0$ and $n \geq 2$

  is a continuous non-linear curve.

The graph of a non-constant (univariate) polynomial always tends to infinity when the variable increases indefinitely (in absolute value).

Polynomial graphs are analyzed in calculus using intercepts, slopes, concavity, and end behavior.

Equations

A *polynomial equation*, also called *algebraic equation*, is an equation of the form

$$a_n x^n + a_{n-1} x^{n-1} + \cdots + a_2 x^2 + a_1 x + a_0 = 0.$$

For example,

$$3x^2 + 4x - 5 = 0.$$

is a polynomial equation.

In case of a univariate polynomial equation, the variable is considered an unknown, and one seeks to find the possible values for which both members of the equation evaluate to the same value (in general more than one solution may exist). A polynomial equation stands in contrast to a *polynomial identity* like $(x + y)(x - y) = x^2 - y^2$, where both expressions represent the same polynomial in different forms, and as a consequence any evaluation of both members gives a valid equality.

In elementary algebra, methods such as the quadratic formula are given for solving all first degree and second degree polynomial equations in one variable. There are also formulas for the cubic and quartic equations. For higher degrees, the Abel–Ruffini theorem asserts that there can not exist a general formula in radicals. However, root-finding algorithms may be used to find numerical approximations of the roots of a polynomial expression of any degree.

The number of real solutions of a polynomial equation with real coefficients may not exceed the degree, and equals the degree when the complex solutions are counted with their multiplicity. This fact is called the fundamental theorem of algebra.

Solving Equations

Every polynomial P in x corresponds to a function, $f(x) = P$ (where the occurrences of x in P are interpreted as the argument of f), called the *polynomial function* of P; the equation in x setting $f(x) = 0$ is the *polynomial equation* corresponding to P. The solutions of this equation are called the *roots* of the polynomial; they are the *zeroes* of the function f (corresponding to the points where the graph of f meets the x-axis). A number a is a root of P if and only if the polynomial $x - a$ (of degree one in x) divides P. It may happen that $x - a$ divides P more than once: if $(x - a)^2$ divides P then a is called a *multiple root* of P, and otherwise a is called a *simple root* of P. If P is a nonzero polynomial, there is a highest power m such that $(x - a)^m$ divides P, which is called the *multiplicity* of the root a in P. When P is the zero polynomial, the corresponding polynomial equation is trivial, and this case is usually excluded when considering roots: with the above definitions every number would be a root of the zero polynomial, with undefined (or infinite) multiplicity. With this exception made, the number of roots of P, even counted with their respective multiplicities, cannot exceed the degree of P. The relation between the roots of a polynomial and its coefficients is described by Vieta's formulas.

Some polynomials, such as $x^2 + 1$, do not have any roots among the real numbers. If, however, the set of allowed candidates is expanded to the complex numbers, every non-constant polynomial has at least one root; this is the fundamental theorem of algebra. By successively dividing out factors $x - a$, one sees that any polynomial with complex coefficients can be written as a constant (its leading coefficient) times a product of such polynomial factors of degree 1; as a consequence, the number of (complex) roots counted with their multiplicities is exactly equal to the degree of the polynomial.

There is a difference between approximating roots and finding exact expressions for roots. Formulas for expressing the roots of polynomials of degree 2 in terms of square roots have been known since ancient times, and for polynomials of degree 3 or 4 similar formulas (using cube roots in addition to square roots) were found in the 16th century (cubic function and quartic function for the formulas and Niccolò Fontana Tartaglia, Lodovico Ferrari, Gerolamo Cardano, and François Viète for historical details). But formulas for degree 5 eluded researchers for several centuries. In 1824, Niels Henrik Abel proved the striking result that there can be no general (finite) formula, involving only arithmetic operations and radicals, that expresses the roots of a polynomial of degree 5 or greater in terms of its coefficients. In 1830, Évariste Galois, studying the permutations of the roots of a polynomial, extended the Abel–Ruffini theorem by showing that, given a polynomial equation, one may decide whether it is solvable by radicals, and, if it is, solve it. This result marked the start of Galois theory and group theory, two important branches of modern mathematics. Galois himself noted that the computations implied by his method were impracticable. Nevertheless, formulas for solvable equations of degrees 5 and 6 have been published.

Numerical approximation of roots of polynomials in one unknown is easily done on a computer by the Jenkins–Traub method, Laguerre's method, Durand–Kerner method, or by some other root-finding algorithm.

For polynomials in more than one indeterminate the notion of root does not exist, and there are usually infinitely many combinations of values for the variables for which the polynomial function takes the value zero. However, for certain *sets* of such polynomials it may happen that for only finitely many combinations all polynomial functions take the value zero.

For a set of polynomial equations in several unknowns, there are algorithms to decide whether they have a finite number of complex solutions. If the number of solutions is finite, there are algorithms to compute the solutions. The methods underlying these algorithms are described in the article System of polynomial equations.

The special case where all the polynomials are of degree one is called a system of linear equations, for which another range of different solution methods exist, including the classical Gaussian elimination.

A polynomial equation for which one is interested only in the solutions which are integers is called a Diophantine equation. Solving Diophantine equations is a very hard task. It has been proved that there cannot be any general algorithm for solving them, and even for deciding whether the set of solutions is empty. Some of the most famous problems that have been solved during the fifty last years are related to Diophantine equations, such as Fermat's Last Theorem.

Generalizations

There are several generalizations of the concept of polynomials.

Trigonometric Polynomials

A trigonometric polynomial is a finite linear combination of functions $\sin(nx)$ and $\cos(nx)$ with n taking on the values of one or more natural numbers. The coefficients may be taken as real numbers, for real-valued functions.

If $\sin(nx)$ and $\cos(nx)$ are expanded in terms of $\sin(x)$ and $\cos(x)$, a trigonometric polynomial becomes a polynomial in the two variables $\sin(x)$ and $\cos(x)$ (using List of trigonometric identities#-Multiple-angle formulae). Conversely, every polynomial in $\sin(x)$ and $\cos(x)$ may be converted, with Product-to-sum identities, into a linear combination of functions sin(nx) and cos(nx). This equivalence explains why linear combinations are called polynomials.

For complex coefficients, there is no difference between such a function and a finite Fourier series.

Trigonometric polynomials are widely used, for example in trigonometric interpolation applied to the interpolation of periodic functions. They are used also in the discrete Fourier transform.

Matrix Polynomials

A matrix polynomial is a polynomial with matrices as variables. Given an ordinary, scalar-valued polynomial:

$$P(x) = \sum_{i=0}^{n} a_i x^i = a_0 + a_1 x + a_2 x^2 + \cdots + a_n x^n,$$

this polynomial evaluated at a matrix A is:

$$P(A) = \sum_{i=0}^{n} a_i A^i = a_0 I + a_1 A + a_2 A^2 + \cdots + a_n A^n,$$

where I is the identity matrix.

A matrix polynomial equation is an equality between two matrix polynomials, which holds for the specific matrices in question. A matrix polynomial identity is a matrix polynomial equation which holds for all matrices A in a specified matrix ring $M_n(R)$.

Laurent Polynomials

Laurent polynomials are like polynomials, but allow negative powers of the variable(s) to occur.

Rational Functions

A rational fraction is the quotient (algebraic fraction) of two polynomials. Any algebraic expression that can be rewritten as a rational fraction is a rational function.

While polynomial functions are defined for all values of the variables, a rational function is defined only for the values of the variables for which the denominator is not zero.

The rational fractions include the Laurent polynomials, but do not limit denominators to powers of an indeterminate.

Power Series

Formal power series are like polynomials, but allow infinitely many non-zero terms to occur, so that they do not have finite degree. Unlike polynomials they cannot in general be explicitly and fully written down (just like irrational numbers cannot), but the rules for manipulating their terms are the same as for polynomials. Non-formal power series also generalize polynomials, but the multiplication of two power series may not converge.

Other Examples

- A bivariate polynomial where the second variable is substituted by an exponential function applied to the first variable, for example $P(x, e^x)$, may be called an exponential polynomial.

Applications

Calculus

The simple structure of polynomial functions makes them quite useful in analyzing general functions using polynomial approximations. An important example in calculus is Taylor's theorem, which roughly states that every differentiable function locally looks like a polynomial function, and the Stone–Weierstrass theorem, which states that every continuous function defined on a compact interval of the real axis can be approximated on the whole interval as closely as desired by a polynomial function.

Calculating derivatives and integrals of polynomial functions is particularly simple. For the polynomial function;

$$\sum_{i=0}^{n} a_i x^i$$

the derivative with respect to x is

$$\sum_{i=1}^{n} a_i i x^{i-1}$$

and the indefinite integral is

$$\sum_{i=0}^{n} \frac{a_i}{i+1} x^{i+1} + c.$$

Abstract Algebra

In abstract algebra, one distinguishes between *polynomials* and *polynomial functions*. A *polynomial f* in one indeterminate x over a ring R is defined as a formal expression of the form:

$$f = a_n x^n + a_{n-1} x^{n-1} + \cdots + a_1 x^1 + a_0 x^0$$

where n is a natural number, the coefficients $a_0, \ldots, a_n$ are elements of R, and x is a formal symbol, whose powers x^i are just placeholders for the corresponding coefficients a_i, so that the given formal expression is just a way to encode the sequence $(a_0, a_1, \ldots)$, where there is an n such that $a_i = 0$ for all $i > n$. Two polynomials sharing the same value of n are considered equal if and only if the sequences of their coefficients are equal; furthermore any polynomial is equal to any polynomial with greater value of n obtained from it by adding terms in front whose coefficient is zero. These polynomials can be added by simply adding corresponding coefficients (the rule for extending by terms with zero coefficients can be used to make sure such coefficients exist). Thus each polynomial is actually equal to the sum of the terms used in its formal expression, if such a term $a_i x^i$ is interpreted as a polynomial that has zero coefficients at all powers of x other than x^i. Then to define multiplication, it suffices by the distributive law to describe the product of any two such terms, which is given by the rule.

$ax^k bx^l = abx^{k+l}$ for all elements a, b of the ring R and all natural numbers k and l.

Thus the set of all polynomials with coefficients in the ring R forms itself a ring, the *ring of polynomials* over R, which is denoted by $R[x]$. The map from R to $R[x]$ sending r to rx^0 is an injective homomorphism of rings, by which R is viewed as a subring of $R[x]$. If R is commutative, then $R[x]$ is an algebra over R.

One can think of the ring $R[x]$ as arising from R by adding one new element x to R, and extending in a minimal way to a ring in which x satisfies no other relations than the obligatory ones, plus commutation with all elements of R (that is $xr = rx$). To do this, one must add all powers of x and their linear combinations as well.

Formation of the polynomial ring, together with forming factor rings by factoring out ideals, are important tools for constructing new rings out of known ones. For instance, the ring (in fact field) of complex numbers, which can be constructed from the polynomial ring $R[x]$ over the real numbers by factoring out the ideal of multiples of the polynomial $x^2 + 1$. Another example is the construction of finite fields, which proceeds similarly, starting out with the field of integers modulo some prime number as the coefficient ring R.

If R is commutative, then one can associate to every polynomial P in $R[x]$, a *polynomial function f* with domain and range equal to R (more generally one can take domain and range to be the same unital associative algebra over R). One obtains the value $f(r)$ by substitution of the value R for the symbol x in P. One reason to distinguish between polynomials and polynomial functions is that over some rings different polynomials may give rise to the same polynomial function. This is not the case when R is the real or complex numbers, whence the two concepts are not always distinguished in analysis. An even more important reason to distinguish between polynomials and polynomial functions is that many operations on polynomials (like Euclidean division) require looking at what a polynomial is composed of as an expression rather than evaluating it at some constant value for x.

Divisibility

In commutative algebra, one major focus of study is *divisibility* among polynomials. If R is an integral domain and f and g are polynomials in $R[x]$, it is said that *f divides g* or f is a divisor of g if there exists a polynomial q in $R[x]$ such that $f q = g$. One can show that every zero gives rise to a linear divisor, or more formally, if f is a polynomial in $R[x]$ and r is an element of R such that $f(r) = 0$, then the polynomial $(x - r)$ divides f. The converse is also true. The quotient can be computed using the polynomial long division.

If F is a field and f and g are polynomials in $F[x]$ with $g \neq 0$, then there exist unique polynomials q and r in $F[x]$ with;

$$f = qg + r$$

and such that the degree of r is smaller than the degree of g (using the convention that the polynomial 0 has a negative degree). The polynomials q and r are uniquely determined by f and g. This is called *Euclidean division, division with remainder* or *polynomial long division* and shows that the ring $F[x]$ is a Euclidean domain.

Analogously, *prime polynomials* (more correctly, *irreducible polynomials*) can be defined as *non-zero polynomials which cannot be factorized into the product of two non constant polynomials.* In the case of coefficients in a ring, *"non constant"* must be replaced by *"non constant or non unit"* (both definitions agree in the case of coefficients in a field). Any polynomial may be decomposed into the product of an invertible constant by a product of irreducible polynomials. If the coefficients belong to a field or a unique factorization domain this decomposition is unique up to the order of the factors and the multiplication of any non unit factor by a unit (and division of the unit factor by the same unit). When the coefficients belong to integers, rational numbers or a finite field, there are algorithms to test irreducibility and to compute the factorization into irreducible polynomials. These algorithms are not practicable for hand written computation, but are available in any computer algebra system. Eisenstein's criterion can also be used in some cases to determine irreducibility.

Other Applications

Polynomials serve to approximate other functions, such as the use of splines.

Polynomials are frequently used to encode information about some other object. The characteristic polynomial of a matrix or linear operator contains information about the operator's eigenvalues. The minimal polynomial of an algebraic element records the simplest algebraic relation satisfied by that element. The chromatic polynomial of a graph counts the number of proper colourings of that graph.

The term "polynomial", as an adjective, can also be used for quantities or functions that can be written in polynomial form. For example, in computational complexity theory the phrase *polynomial time* means that the time it takes to complete an algorithm is bounded by a polynomial function of some variable, such as the size of the input.

History

Determining the roots of polynomials, or "solving algebraic equations", is among the oldest problems in mathematics. However, the elegant and practical notation we use today only developed beginning in the 15th century. Before that, equations were written out in words. For example, an algebra problem from the Chinese Arithmetic in Nine Sections, circa 200 BCE, begins "Three sheafs of good crop, two sheafs of mediocre crop, and one sheaf of bad crop are sold for 29 dou." We would write $3x + 2y + z = 29$.

History of the Notation

The earliest known use of the equal sign is in Robert Recorde's *The Whetstone of Witte*, 1557. The signs + for addition, − for subtraction, and the use of a letter for an unknown appear in Michael Stifel's *Arithemetica integra*, 1544. René Descartes, in *La géometrie*, 1637, introduced the concept of the graph of a polynomial equation. He popularized the use of letters from the beginning of the alphabet to denote constants and letters from the end of the alphabet to denote variables, as can be seen above, in the general formula for a polynomial in one variable, where the a's denote constants and x denotes a variable. Descartes introduced the use of superscripts to denote exponents as well.

Differential Equation

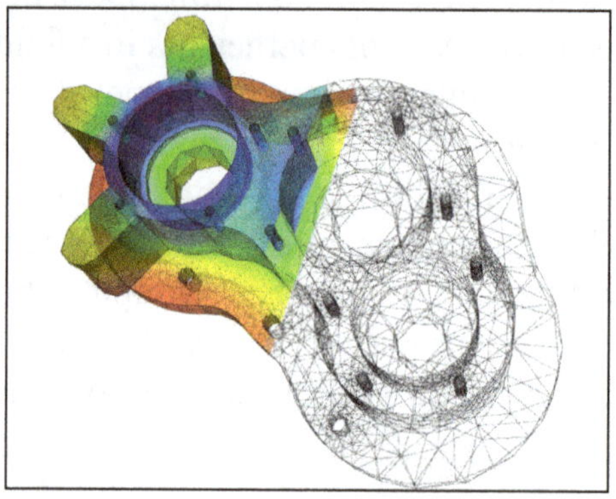

Visualization of heat transfer in a pump casing, created by solving the heat equation. Heat is being generated internally in the casing and being cooled at the boundary, providing a steady state temperature distribution.

A differential equation is a mathematical equation that relates some function with its derivatives. In applications, the functions usually represent physical quantities, the derivatives represent their rates of change, and the equation defines a relationship between the two. Because such relations are extremely common, differential equations play a prominent role in many disciplines including engineering, physics, economics, and biology.

In pure mathematics, differential equations are studied from several different perspectives, mostly concerned with their solutions—the set of functions that satisfy the equation. Only the simplest differential equations are solvable by explicit formulas; however, some properties of solutions of a given differential equation may be determined without finding their exact form.

If a self-contained formula for the solution is not available, the solution may be numerically approximated using computers. The theory of dynamical systems puts emphasis on qualitative analysis of systems described by differential equations, while many numerical methods have been developed to determine solutions with a given degree of accuracy.

History

Differential equations first came into existence with the invention of calculus by Newton and Leibniz. In Chapter 2 of his 1671 work "Methodus fluxionum et Serierum Infinitarum", Isaac Newton listed three kinds of differential equations:

$$\frac{dy}{dx} = f(x)$$

$$\frac{dy}{dx} = f(x, y)$$

$$x_1 \frac{\partial y}{\partial x_1} + x_2 \frac{\partial y}{\partial x_2} = y$$

He solves these examples and others using infinite series and discusses the non-uniqueness of solutions.

Jacob Bernoulli proposed the Bernoulli differential equation in 1695. This is an ordinary differential equation of the form;

$$y' + P(x)y = Q(x)y^n,$$

for which the following year Leibniz obtained solutions by simplifying it.

Historically, the problem of a vibrating string such as that of a musical instrument was studied by Jean le Rond d'Alembert, Leonhard Euler, Daniel Bernoulli, and Joseph-Louis Lagrange. In 1746, d'Alembert discovered the one-dimensional wave equation, and within ten years Euler discovered the three-dimensional wave equation.

The Euler–Lagrange equation was developed in the 1750s by Euler and Lagrange in connection with their studies of the tautochrone problem. This is the problem of determining a curve on which a weighted particle will fall to a fixed point in a fixed amount of time, independent of the starting point.

Lagrange solved this problem in 1755 and sent the solution to Euler. Both further developed Lagrange's method and applied it to mechanics, which led to the formulation of Lagrangian mechanics.

Fourier published his work on heat flow in *Théorie analytique de la chaleur* (The Analytic Theory of Heat), in which he based his reasoning on Newton's law of cooling, namely, that the flow of heat between two adjacent molecules is proportional to the extremely small difference of their temperatures. Contained in this book was Fourier's proposal of his heat equation for conductive diffusion of heat. This partial differential equation is now taught to every student of mathematical physics.

Example

For example, in classical mechanics, the motion of a body is described by its position and velocity as the time value varies. Newton's laws allow (given the position, velocity, acceleration and various forces acting on the body) one to express these variables dynamically as a differential equation for the unknown position of the body as a function of time.

In some cases, this differential equation (called an equation of motion) may be solved explicitly.

An example of modelling a real world problem using differential equations is the determination of the velocity of a ball falling through the air, considering only gravity and air resistance. The ball's acceleration towards the ground is the acceleration due to gravity minus the acceleration due to air resistance.

Gravity is considered constant, and air resistance may be modeled as proportional to the ball's velocity. This means that the ball's acceleration, which is a derivative of its velocity, depends on the velocity (and the velocity depends on time). Finding the velocity as a function of time involves solving a differential equation and verifying its validity.

Types

Differential equations can be divided into several types. Apart from describing the properties of the equation itself, these classes of differential equations can help inform the choice of approach to a solution. Commonly used distinctions include whether the equation is: Ordinary/Partial, Linear/Non-linear, and Homogeneous/Inhomogeneous. This list is far from exhaustive; there are many other properties and subclasses of differential equations which can be very useful in specific contexts.

Ordinary Differential Equations

An ordinary differential equation (*ODE*) is an equation containing a function of one independent variable and its derivatives. The term "*ordinary*" is used in contrast with the term partial differential equation which may be with respect to *more than* one independent variable.

Linear differential equations, which have solutions that can be added and multiplied by coefficients, are well-defined and understood, and exact closed-form solutions are obtained. By contrast, ODEs that lack additive solutions are nonlinear, and solving them is far more intricate, as one can rarely represent them by elementary functions in closed form: Instead, exact and analytic solutions of ODEs are in series or integral form. Graphical and numerical methods, applied by hand or by computer, may approximate solutions of ODEs and perhaps yield useful information, often sufficing in the absence of exact, analytic solutions.

Partial Differential Equations

A partial differential equation (*PDE*) is a differential equation that contains unknown multivariable functions and their partial derivatives. (This is in contrast to ordinary differential equations, which deal with functions of a single variable and their derivatives.) PDEs are used to formulate problems involving functions of several variables, and are either solved in closed form, or used to create a relevant computer model.

PDEs can be used to describe a wide variety of phenomena such as sound, heat, electrostatics, electrodynamics, fluid flow, elasticity, or quantum mechanics. These seemingly distinct physical phenomena can be formalised similarly in terms of PDEs. Just as ordinary differential equations often model one-dimensional dynamical systems, partial differential equations often model multidimensional systems. PDEs find their generalisation in stochastic partial differential equations.

Linear Differential Equations

A differential equation is *linear* if the unknown function and its derivatives have *degree* 1 (products of the unknown function and its derivatives are not allowed) and *nonlinear* otherwise. The characteristic property of linear equations is that their solutions form an affine subspace of an appropriate function space, which results in much more developed theory of linear differential equations.

Homogeneous linear differential equations are a subclass of linear differential equations for which the space of solutions is a linear subspace i.e. the sum of any set of solutions or multiples of solutions is also a solution. The coefficients of the unknown function and its derivatives in a linear differential equation are allowed to be (known) functions of the independent variable or variables; if these coefficients are constants then one speaks of a *constant coefficient linear differential equation.*

Non-linear Differential Equations

Non-linear differential equations are formed by the *products of the unknown function and its derivatives* are allowed and its degree is > 1.There are very few methods of solving nonlinear differential equations exactly; those that are known typically depend on the equation having particular symmetries. Nonlinear differential equations can exhibit very complicated behavior over extended time intervals, characteristic of chaos. Even the fundamental questions of existence, uniqueness, and extendability of solutions for nonlinear differential equations, and well-posedness of initial and boundary value problems for nonlinear PDEs are hard problems and their resolution in special cases is considered to be a significant advance in the mathematical theory (cf. Navier–Stokes existence and smoothness). However, if the differential equation is a correctly formulated representation of a meaningful physical process, then one expects it to have a solution.

Linear differential equations frequently appear as approximations to nonlinear equations. These approximations are only valid under restricted conditions. For example, the harmonic oscillator equation is an approximation to the nonlinear pendulum equation that is valid for small amplitude oscillations.

Equation Order

Differential equations are described by their order, determined by the term with the highest derivatives. An equation containing only first derivatives is a *first-order differential equation*, an equation containing the second derivative is a *second-order differential equation*, and so on.

Examples

In the first group of examples, let u be an unknown function of x, and c and ω are known constants. Note both ordinary and partial differential equations are broadly classified as *linear* and *nonlinear*.

- Inhomogeneous first-order linear constant coefficient ordinary differential equation:

$$\frac{du}{dx} = cu + x^2.$$

- Homogeneous second-order linear ordinary differential equation:

$$\frac{d^2u}{dx^2} - x\frac{du}{dx} + u = 0.$$

- Homogeneous second-order linear constant coefficient ordinary differential equation describing the harmonic oscillator:

$$\frac{d^2u}{dx^2} + \omega^2 u = 0.$$

- Inhomogeneous first-order nonlinear ordinary differential equation:

$$\frac{du}{dx} = u^2 + 4.$$

- Second-order nonlinear (due to sine function) ordinary differential equation describing the motion of a pendulum of length *L*:

$$L\frac{d^2u}{dx^2} + g\sin u = 0.$$

In the next group of examples, the unknown function *u* depends on two variables *x* and *t* or *x* and *y*.

- Homogeneous first-order linear partial differential equation:

$$\frac{\partial u}{\partial t} + t\frac{\partial u}{\partial x} = 0.$$

- Homogeneous second-order linear constant coefficient partial differential equation of elliptic type, the Laplace equation:

$$\frac{\partial^2 u}{\partial x^2} + \frac{\partial^2 u}{\partial y^2} = 0.$$

$$\frac{\partial u}{\partial t} = 6u\frac{\partial u}{\partial x} - \frac{\partial^3 u}{\partial x^3}.$$

Existence of Solutions

Solving differential equations is not like solving algebraic equations. Not only are their solutions oftentimes unclear, but whether solutions are unique or exist at all are also notable subjects of interest.

For first order initial value problems, the Peano existence theorem gives one set of circumstances in which a solution exists. Given any point (a,b) in the xy-plane, define some rectangular region Z, such that $Z = [l,m] \times [n,p]$ and (a,b) is in the interior of Z. If we are given a differential equation $\frac{dy}{dx} = g(x,y)$ and the condition that $y = b$ when $x = a$, then there is locally a solution to this problem if $g(x,y)$ and $\frac{\partial g}{\partial x}$ are both continuous on Z. This solution exists on some interval with its center at a. The solution may not be unique.

However, this only helps us with first order initial value problems. Suppose we had a linear initial value problem of the nth order:

$$f_n(x)\frac{d^n y}{dx^n} + \cdots + f_1(x)\frac{dy}{dx} + f_0(x)y = g(x).$$

such that

$$y(x_0) = y_0, y'(x_0) = y_0', y''(x_0) = y_0'', \cdots$$

For any nonzero $f_n(x)$, if $\{f_0, f_1, \cdots\}$ and g are continuous on some interval containing x_0, y is unique and exists.

Related Concepts

- A delay differential equation (DDE) is an equation for a function of a single variable, usually called time, in which the derivative of the function at a certain time is given in terms of the values of the function at earlier times.
- A stochastic differential equation (SDE) is an equation in which the unknown quantity is a stochastic process and the equation involves some known stochastic processes, for example, the Wiener process in the case of diffusion equations.
- A differential algebraic equation (DAE) is a differential equation comprising differential and algebraic terms, given in implicit form.

Connection to Difference Equations

The theory of differential equations is closely related to the theory of difference equations, in which the coordinates assume only discrete values, and the relationship involves values of the unknown function or functions and values at nearby coordinates. Many methods to compute numerical solutions of differential equations or study the properties of differential equations involve approximation of the solution of a differential equation by the solution of a corresponding difference equation.

Applications

The study of differential equations is a wide field in pure and applied mathematics, physics, and engineering. All of these disciplines are concerned with the properties of differential equations of various types. Pure mathematics focuses on the existence and uniqueness of solutions, while applied mathematics emphasizes the rigorous justification of the methods for approximating solutions. Differential equations play an important role in modelling virtually every physical, technical, or biological process, from celestial motion, to bridge design, to interactions between neurons. Differential equations such as those used to solve real-life problems may not necessarily be directly solvable, i.e. do not have closed form solutions. Instead, solutions can be approximated using numerical methods.

Many fundamental laws of physics and chemistry can be formulated as differential equations. In biology and economics, differential equations are used to model the behavior of complex systems. The mathematical theory of differential equations first developed together with the sciences where the equations had originated and where the results found application. However, diverse problems, sometimes originating in quite distinct scientific fields, may give rise to identical differential equations. Whenever this happens, mathematical theory behind the equations can be viewed as a unifying principle behind diverse phenomena. As an example, consider propagation of light and sound in the atmosphere, and of waves on the surface of a pond. All of them may be described by the same second-order partial differential equation, the wave equation, which allows us to think of light and sound as forms of waves, much like familiar waves in the water. Conduction of heat, the theory of which was developed by Joseph Fourier, is governed by another second-order partial differential equation, the heat equation. It turns out that many diffusion processes, while seemingly different, are described by the same equation; the Black–Scholes equation in finance is, for instance, related to the heat equation.

Physics

- Euler–Lagrange equation in classical mechanics
- Hamilton's equations in classical mechanics
- Radioactive decay in nuclear physics
- Newton's law of cooling in thermodynamics
- The wave equation
- The heat equation in thermodynamics
- Laplace's equation, which defines harmonic functions
- Poisson's equation
- The geodesic equation
- The Navier–Stokes equations in fluid dynamics
- The Diffusion equation in stochastic processes
- The Convection–diffusion equation in fluid dynamics
- The Cauchy–Riemann equations in complex analysis
- The Poisson–Boltzmann equation in molecular dynamics
- The shallow water equations
- Universal differential equation
- The Lorenz equations whose solutions exhibit chaotic flow.

Classical Mechanics

So long as the force acting on a particle is known, Newton's second law is sufficient to describe the motion of a particle. Once independent relations for each force acting on a particle are available, they can be substituted into Newton's second law to obtain an ordinary differential equation, which is called the *equation of motion.*

Electrodynamics

Maxwell's equations are a set of partial differential equations that, together with the Lorentz force law, form the foundation of classical electrodynamics, classical optics, and electric circuits. These fields in turn underlie modern electrical and communications technologies. Maxwell's equations describe how electric and magnetic fields are generated and altered by each other and by charges and currents. They are named after the Scottish physicist and mathematician James Clerk Maxwell, who published an early form of those equations between 1861 and 1862.

General Relativity

The Einstein field equations (EFE; also known as "Einstein's equations") are a set of ten partial differential equations in Albert Einstein's general theory of relativity which describe the fundamental interaction of gravitation as a result of spacetime being curved by matter and energy. First published by Einstein in 1915 as a tensor equation, the EFE equate local spacetime curvature (expressed by the Einstein tensor) with the local energy and momentum within that spacetime (expressed by the stress–energy tensor).

Quantum Mechanics

In quantum mechanics, the analogue of Newton's law is Schrödinger's equation (a partial differential equation) for a quantum system (usually atoms, molecules, and subatomic particles whether free, bound, or localized). It is not a simple algebraic equation, but in general a linear partial differential equation, describing the time-evolution of the system's wave function (also called a "state function").

Biology

- Verhulst equation – biological population growth
- von Bertalanffy model – biological individual growth
- Replicator dynamics – found in theoretical biology
- Hodgkin–Huxley model – neural action potentials

Predator-prey Equations

The Lotka–Volterra equations, also known as the predator–prey equations, are a pair of first-order, non-linear, differential equations frequently used to describe the dynamics of biological systems in which two species interact, one as a predator and the other as prey.

Chemistry

The *rate law* or rate equation for a chemical reaction is a differential equation that links the reaction rate with concentrations or pressures of reactants and constant parameters (normally rate coefficients and partial reaction orders). To determine the rate equation for a particular system one combines the reaction rate with a mass balance for the system.

Economics

- The key equation of the Solow–Swan model is $\frac{\partial k(t)}{\partial t} = s[k(t)]^{\alpha} - \delta k(t)$

- The Black–Scholes PDE
- Malthusian growth model
- The Vidale–Wolfe advertising model

Integral Equation

In mathematics, an integral equation is an equation in which an unknown function appears under an integral sign.

There is a close connection between differential and integral equations, and some problems may be formulated either way.

Overview

The most basic type of integral equation is called a *Fredholm equation of the first type*,

$$f(x) \quad \int K(x,t) \quad (t)dt.$$

The notation follows Arfken. Here φ is an unknown function, f is a known function, and K is another known function of two variables, often called the kernel function. Note that the limits of integration are constant: this is what characterizes a Fredholm equation.

If the unknown function occurs both inside and outside of the integral, the equation is known as a *Fredholm equation of the second type*,

$$\varphi(x) = f(x) + \lambda \int_a^b K(x,t)\varphi(t)dt.$$

The parameter λ is an unknown factor, which plays the same role as the eigenvalue in linear algebra.

If one limit of integration is a variable, the equation is called a Volterra equation. The following are called *Volterra equations of the first and second types*, respectively,

$$f(x) = \int_a^x K(x,t)\varphi(t)dt$$

$$\varphi(x) = f(x) + \lambda \int_a^x K(x,t)\varphi(t)dt.$$

In all of the above, if the known function f is identically zero, the equation is called a *homogeneous integral equation*. If f is nonzero, it is called an *inhomogeneous integral equation*.

Numerical Solution

It is worth noting that integral equations often do not have an analytical solution, and must be solved numerically. An example of this is evaluating the Electric-Field Integral Equation (EFIE) or Magnetic-Field Integral Equation (MFIE) over an arbitrarily shaped object in an electromagnetic scattering problem.

One method to solve numerically requires discretizing variables and replacing integral by a quadrature rule;

$$\sum_{j=1}^{n} w_j K\left(s_i, t_j\right) u(t_j) = f(s_i), \qquad i = 0, 1, \cdots, n.$$

Then we have a system with n equations and n variables. By solving it we get the value of the n variables;

$$u(t_0), u(t_1), \cdots, u(t_n).$$

Classification

Integral equations are classified according to three different dichotomies, creating eight different kinds:

Limits of integration

- both fixed: Fredholm equation
- one variable: Volterra equation

Placement of unknown function

- only inside integral: first kind
- both inside and outside integral: second kind

Nature of known function f

- identically zero: homogeneous
- not identically zero: inhomogeneous

Integral equations are important in many applications. Problems in which integral equations are encountered include radiative transfer, and the oscillation of a string, membrane, or axle. Oscillation problems may also be solved as differential equations.

Both Fredholm and Volterra equations are linear integral equations, due to the linear behaviour of $\varphi(x)$ under the integral. A nonlinear Volterra integral equation has the general form:

$$\varphi(x) = f(x) + \lambda \int_a^x K(x,t) F(x,t,\varphi(t))\,dt,$$

where F is a known function.

Wiener–Hopf Integral Equations

$$y(t) = \lambda x(t) + \int_0^\infty k(t-s) x(s)\,ds, \qquad 0 \le t < \infty.$$

Originally, such equations were studied in connection with problems in radiative transfer, and more recently, they have been related to the solution of boundary integral equations for planar problems in which the boundary is only piecewise smooth.

Power Series Solution for Integral Equations

In many cases, if the Kernel of the integral equation is of the form $K(xt)$ and the Mellin transform of $K(t)$ exists, we can find the solution of the integral equation;

$$g(s) = s\int_0^\infty dt K(st) f(t)$$

in the form of a power series

$$f(t) = \sum_{n=0}^{\infty} \frac{a_n}{M(n+1)} t^n$$

where

$$g(s) = \sum_{n=0}^{\infty} a_n s^{-n}, \qquad M(n+1) = \int_0^\infty dt K(t) t^n$$

are the Z-transform of the function $g(s)$, and $M(n + 1)$ is the Mellin transform of the Kernel.

Integral Equations as a Generalization of Eigenvalue Equations

Certain homogeneous linear integral equations can be viewed as the continuum limit of eigenvalue equations. Using index notation, an eigenvalue equation can be written as;

$$\sum_j M_{i,j} v_j = \lambda v_i$$

where M = $[M_{i,j}]$ is a matrix, v is one of its eigenvectors, and λ is the associated eigenvalue.

Taking the continuum limit, i.e., replacing the discrete indices i and j with continuous variables x and y, yields

$$\int K(x, y)\varphi(y)\mathrm{d}y = \lambda\varphi(x),$$

where the sum over j has been replaced by an integral over y and the matrix M and the vector v have been replaced by the *kernel* $K(x, y)$ and the eigenfunction $\varphi(y)$. (The limits on the integral are fixed, analogously to the limits on the sum over j.) This gives a linear homogeneous Fredholm equation of the second type.

In general, $K(x, y)$ can be a distribution, rather than a function in the strict sense. If the distribution K has support only at the point $x = y$, then the integral equation reduces to a differential eigenfunction equation.

In general, Volterra and Fredholm integral equations can arise from a single differential equation, depending on which sort of conditions are applied at the boundary of the domain of its solution.

Diophantine Equation

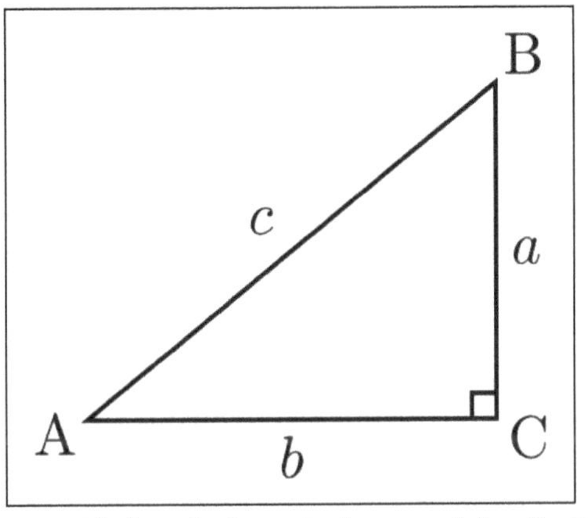

Finding all right triangles with integer side-lengths is equivalent to solving the Diophantine equation $a^2 + b^2 = c^2$.

In mathematics, a Diophantine equation is a polynomial equation, usually in two or more unknowns, such that only the integer solutions are sought or studied (an integer solution is a solution such that all the unknowns take integer values). A linear Diophantine equation is an equation between two sums of monomials of degree zero or one. An exponential Diophantine equation is one in which exponents on terms can be unknowns.

Diophantine problems have fewer equations than unknown variables and involve finding integers that work correctly for all equations. In more technical language, they define an algebraic curve, algebraic surface, or more general object, and ask about the lattice points on it.

The word *Diophantine* refers to the Hellenistic mathematician of the 3rd century, Diophantus of Alexandria, who made a study of such equations and was one of the first mathematicians to introduce symbolism into algebra. The mathematical study of Diophantine problems that Diophantus initiated is now called Diophantine analysis.

While individual equations present a kind of puzzle and have been considered throughout history, the formulation of general theories of Diophantine equations (beyond the theory of quadratic forms) was an achievement of the twentieth century.

Examples

In the following Diophantine equations, w, x, y, and z are the unknowns and the other letters are given constants:

$ax + by = 1$	This is a linear Diophantine equation.

$w^3 + x^3 = y^3 + z^3$	The smallest nontrivial solution in positive integers is $12^3 + 1^3 = 9^3 + 10^3 = 1729$. It was famously given as an evident property of 1729, a taxicab number (also named Hardy–Ramanujan number) by Ramanujan to Hardy while meeting in 1917. There are infinitely many nontrivial solutions.
$x^n + y^n = z^n$	For $n = 2$ there are infinitely many solutions (x,y,z): the Pythagorean triples. For larger integer values of n, Fermat's Last Theorem (initially claimed in 1637 by Fermat and proved by Wiles in 1995) states there are no positive integer solutions (x,y,z).
$x^2 - ny^2 = \pm 1$	This is Pell's equation, which is named after the English mathematician John Pell. It was studied by Brahmagupta in the 7th century, as well as by Fermat in the 17th century.
$4/n = 1/x + 1/y + 1/z$	The Erdős–Straus conjecture states that, for every positive integer $n \geq 2$, there exists a solution in x, y, and z, all as positive integers. Although not usually stated in polynomial form, this example is equivalent to the polynomial equation $4xyz = yzn + xzn + xyn = n(yz + xz + xy)$.
$x^4 + y^4 + z^4 = w^4$	Conjectured incorrectly by Euler to have no nontrivial solutions. Proved by Elkies to have infinitely many nontrivial solutions, with a computer search by Frye determining the smallest nontrivial solution.

Linear Diophantine Equations

One Equation

The simplest linear Diophantine equation takes the form $ax + by = c$, where a, b and c are given integers. The solutions are described by the following theorem:

> *This Diophantine equation has a solution* (where x and y are integers) *if and only if c is a multiple of the greatest common divisor of a and b. Moreover, if (x, y) is a solution, then the other solutions have the form $(x + kv, y - ku)$, where k is an arbitrary integer, and u and v are the quotients of a and b (respectively) by the greatest common divisor of a and b.*

Proof: If d is this greatest common divisor, Bézout's identity asserts the existence of integers e and f such that $ae + bf = d$. If c is a multiple of d, then $c = dh$ for some integer h, and (eh, fh) is a solution. On the other hand, for every pair of integers x and y, the greatest common divisor d of a and b divides $ax + by$. Thus, if the equation has a solution, then c must be a multiple of d. If $a = ud$ and $b = vd$, then for every solution (x, y), we have.

$$a(x + kv) + b(y - ku) = ax + by + k(av - bu) = ax + by + k(udv - vdu) = ax + by,$$

showing that $(x + kv, y - ku)$ is another solution. Finally, given two solutions such that $ax_1 + by_1 = ax_2 + by_2 = c$, one deduces that $u(x_2 - x_1) + v(y_2 - y_1) = 0$. As u and v are coprime, Euclid's lemma shows that there exists an integer k such that $x_2 - x_1 = kv$ and $y_2 - y_1 = -ku$. Therefore, $x_2 = x_1 + kv$ and $y_2 = y_1 - ku$, which completes the proof.

Chinese Remainder Theorem

The Chinese remainder theorem describes an important class of linear Diophantine systems of equations: let $n_1, ..., n_k$ be k pairwise coprime integers greater than one, $a_1, ..., a_k$ be k arbitrary integers, and N be the product $n_1 \cdots n_k$. The Chinese remainder theorem asserts that the following linear Diophantine system has exactly one solution $(x, x_1, ..., x_k)$ such that $0 \leq x < N$, and that the other solutions are obtained by adding to x a multiple of N:

$$x = a_1 + n_1 x_1$$
$$\vdots$$
$$x = a_k + n_k x_k$$

System of Linear Diophantine Equations

More generally, every system of linear Diophantine equations may be solved by computing the Smith normal form of its matrix, in a way that is similar to the use of the reduced row echelon form to solve a system of linear equations over a field. Using matrix notation every system of linear Diophantine equations may be written.

$$AX = C,$$

where A is an $m \times n$ matrix of integers, X is an $n \times 1$ column matrix of unknowns and C is an $m \times 1$ column matrix of integers.

The computation of the Smith normal form of A provides two unimodular matrices (that is matrices that are invertible over the integers and have ±1 as determinant) U and V of respective dimensions $m \times m$ and $n \times n$, such that the matrix.

$$B = [b_{i,j}] = UAV$$

is such that $b_{i,i}$ is not zero for i not greater than some integer k, and all the other entries are zero. The system to be solved may thus be rewritten as.

$$B(V^{-1}X) = UC.$$

Calling y_i the entries of $V^{-1}X$ and d_i those of $D = UC$, this leads to the system

$$b_{i,i}y_i = d_i \text{ for } 1 \le i \le k,$$

$$0y_i = d_i \text{ for } k < i \le n.$$

This system is equivalent to the given one in the following sense: A column matrix of integers x is a solution of the given system if and only if $x = Vy$ for some column matrix of integers y such that $By = D$.

It follows that the system has a solution if and only if $b_{i,i}$ divides d_i for $i \le k$ and $d_i = 0$ for $i > k$. If this condition is fulfilled, the solutions of the given system are:

$$V\begin{bmatrix} \frac{d_1}{b_{1,1}} \\ \vdots \\ \frac{d_k}{b_{k,k}} \\ h_{k+1} \\ \vdots \\ h_n \end{bmatrix},$$

where $h_{k+1}, \ldots, h_n$ are arbitrary integers.

Hermite normal form may also be used for solving systems of linear Diophantine equations. However, Hermite normal form does not directly provide the solutions; to get the solutions from the Hermite normal form, one has to successively solve several linear equations. Nevertheless, Richard Zippel wrote that the Smith normal form "is somewhat more than is actually needed to solve linear diophantine equations. Instead of reducing the equation to diagonal form, we only need to make it triangular, which is called the Hermite normal form. The Hermite normal form is substantially easier to compute than the Smith normal form."

Integer linear programming amounts to finding some integer solutions (optimal in some sense) of linear systems that include also inequations. Thus systems of linear Diophantine equations are basic in this context, and textbooks on integer programming usually have a treatment of systems of linear Diophantine equations.

Diophantine Analysis

Typical Questions

The questions asked in Diophantine analysis include:

1. Are there any solutions?
2. Are there any solutions beyond some that are easily found by inspection?
3. Are there finitely or infinitely many solutions?
4. Can all solutions be found in theory?
5. Can one in practice compute a full list of solutions?

These traditional problems often lay unsolved for centuries, and mathematicians gradually came to understand their depth (in some cases), rather than treat them as puzzles.

Typical Problem

The given information is that a father's age is 1 less than twice that of his son, and that the digits AB making up the father's age are reversed in the son's age (i.e. BA). This leads to the equation $10A + B = 2(10B + A) - 1$, thus $19B - 8A = 1$. Inspection gives the result $A = 7$, $B = 3$, and thus AB equals 73 years and BA equals 37 years. One may easily show that there is not any other solution with A and B positive integers less than 10.

17th and 18th Centuries

In 1637, Pierre de Fermat scribbled on the margin of his copy of *Arithmetica*: "It is impossible to separate a cube into two cubes, or a fourth power into two fourth powers, or in general, any power higher than the second into two like powers." Stated in more modern language, "The equation $a^n + b^n = c^n$ has no solutions for any n higher than 2." And then he wrote, intriguingly: "I have discovered a truly marvelous proof of this proposition, which this margin is too narrow to contain." Such a proof eluded mathematicians for centuries, however, and as such his statement became famous as Fermat's Last Theorem. It wasn't until 1995 that it was proven by the British mathematician Andrew Wiles.

In 1657, Fermat attempted to solve the Diophantine equation $61x^2 + 1 = y^2$ (solved by Brahmagupta over 1000 years earlier). The equation was eventually solved by Euler in the early 18th century, who also solved a number of other Diophantine equations. The smallest solution of this equation in positive integers is $x = 226153980$, $y = 1766319049$.

Hilbert's Tenth Problem

In 1900, David Hilbert proposed the solvability of all Diophantine equations as the tenth of his fundamental problems. In 1970, Yuri Matiyasevich solved it negatively, by proving that a general algorithm for solving all Diophantine equations cannot exist.

Diophantine Geometry

Diophantine geometry, which is the application of techniques from algebraic geometry in this field, has continued to grow as a result; since treating arbitrary equations is a dead end, attention turns to equations that also have a geometric meaning. The central idea of Diophantine geometry is that of a rational point, namely a solution to a polynomial equation or a system of polynomial equations, which is a vector in a prescribed field K, when K is *not* algebraically closed.

Modern Research

One of the few general approaches is through the Hasse principle. Infinite descent is the traditional method, and has been pushed a long way.

The depth of the study of general Diophantine equations is shown by the characterisation of Diophantine sets as equivalently described as recursively enumerable. In other words, the general problem of Diophantine analysis is blessed or cursed with universality, and in any case is not something that will be solved except by re-expressing it in other terms.

The field of Diophantine approximation deals with the cases of *Diophantine inequalities*. Here variables are still supposed to be integral, but some coefficients may be irrational numbers, and the equality sign is replaced by upper and lower bounds.

The most celebrated single question in the field, the conjecture known as Fermat's Last Theorem, was solved by Andrew Wiles but using tools from algebraic geometry developed during the last century rather than within number theory where the conjecture was originally formulated. Other major results, such as Faltings' theorem, have disposed of old conjectures.

Infinite Diophantine Equations

An example of an infinite diophantine equation is:

$$n = a^2 + 2b^2 + 3c^2 + 4d^2 + 5e^2 + \dots,$$

which can be expressed as "How many ways can a given integer n be written as the sum of a square plus twice a square plus thrice a square and so on?" The number of ways this can be done for each n forms an integer sequence. Infinite Diophantine equations are related to theta functions and infinite dimensional lattices. This equation always has a solution for any positive n. Compare this to:

$$n = a^2 + 4b^2 + 9c^2 + 16d^2 + 25e^2 + ...,$$

which does not always have a solution for positive n.

Exponential Diophantine Equations

If a Diophantine equation has as an additional variable or variables occurring as exponents, it is an exponential Diophantine equation. Examples include the Ramanujan–Nagell equation, $2^n - 7 = x^2$, and the equation of the Fermat-Catalan conjecture and Beal's conjecture, $a^m + b^n = c^k$ with inequality restrictions on the exponents. A general theory for such equations is not available; particular cases such as Catalan's conjecture have been tackled. However, the majority are solved via ad hoc methods such as Størmer's theorem or even trial and error.

References

- Shorey, T. N.; Tijdeman, R. (1986). Exponential Diophantine equations. Cambridge Tracts in Mathematics. 87. Cambridge University Press. ISBN 0-521-26826-5. Zbl 0606.10011
- Misner, Charles W.; Thorne, Kip S.; Wheeler, John Archibald. Gravitation. San Francisco: W. H. Freeman. ISBN 978-0-7167-0344-0 Chapter 34, p. 916
- Hairer, Ernst; Nørsett, Syvert Paul; Wanner, Gerhard (1993), Solving ordinary differential equations I: Nonstiff problems, Berlin, New York: Springer-Verlag, ISBN 978-3-540-56670-0
- Lang, Serge (2002), Algebra, Graduate Texts in Mathematics, 211 (Revised third ed.), New York: Springer-Verlag, ISBN 978-0-387-95385-4, MR 1878556. This classical book covers most of the content of this article
- Sethuraman, B.A. (1997). "Polynomials". Rings, Fields, and Vector Spaces: An Introduction to Abstract Algebra Via Geometric Constructibility. Springer. ISBN 978-0-387-94848-5
- Cahen, Paul-Jean; Chabert, Jean-Luc (1997). Integer-Valued Polynomials. American Mathematical Society. ISBN 978-0-8218-0388-2
- Bronstein, Manuel; et al., eds. (2006). Solving Polynomial Equations: Foundations, Algorithms, and Applications. Springer. ISBN 978-3-540-27357-8
- Smart, Nigel P. (1998). The algorithmic resolution of Diophantine equations. London Mathematical Society Student Texts. 41. Cambridge University Press. ISBN 0-521-64156-X. Zbl 0907.11001
- The Cambridge Ancient History Part 2 Early History of the Middle East. Cambridge University Press. 1971. p. 530. ISBN 978-0-521-07791-0
- Schmidt, Wolfgang M. (1991). Diophantine approximations and Diophantine equations. Lecture Notes in Mathematics. 1467. Berlin: Springer-Verlag. ISBN 3-540-54058-X. Zbl 0754.11020
- Barnett, R.A.; Ziegler, M.R.; Byleen, K.E. (2008), College Mathematics for Business, Economics, Life Sciences and the Social Sciences (11th ed.), Upper Saddle River, N.J.: Pearson, ISBN 0-13-157225-3
- Andrei D. Polyanin and Alexander V. Manzhirov Handbook of Integral Equations. CRC Press, Boca Raton, 1998. ISBN 0-8493-2876-4

- Rich, Barnett; Schmidt, Philip (2004), Schaum's Outline of Theory and Problems of Elementary Algebra, The McGraw–Hill Companies, ISBN 0-07-141083-X, Chapter 13 §4.4, p. 291
- Stillwell, John (2004). Mathematics and its History (Second ed.). Springer Science + Business Media Inc. ISBN 0-387-95336-1

5

Geometric Algebra

Geometric Algebra deals with the vector space with quadratic form and geometric product. Algebra elements called multivectors, contains scalars and vectors. Geometric Algebra is a vast subject that branches out into significant sub-disciplines. Topics such as Algebraic variety, Hilbert's Nullstellensatz and Bézout's theorem have been thoroughly discussed in this chapter.

Algebraic Geometry

Algebraic geometry is a branch of mathematics, classically studying zeros of multivariate polynomials. Modern algebraic geometry is based on the use of abstract algebraic techniques, mainly from commutative algebra, for solving geometrical problems about these sets of zeros.

The fundamental objects of study in algebraic geometry are algebraic varieties, which are geometric manifestations of solutions of systems of polynomial equations. Examples of the most studied classes of algebraic varieties are: plane algebraic curves, which include lines, circles, parabolas, ellipses, hyperbolas, cubic curves like elliptic curves, and quartic curves like lemniscates and Cassini ovals. A point of the plane belongs to an algebraic curve if its coordinates satisfy a given polynomial equation. Basic questions involve the study of the points of special interest like the singular points, the inflection points and the points at infinity. More advanced questions involve the topology of the curve and relations between the curves given by different equations.

Algebraic geometry occupies a central place in modern mathematics and has multiple conceptual connections with such diverse fields as complex analysis, topology and number theory. Initially a study of systems of polynomial equations in several variables, the subject of algebraic geometry starts where equation solving leaves off, and it becomes even more important to understand the intrinsic properties of the totality of solutions of a system of equations, than to find a specific solution; this leads into some of the deepest areas in all of mathematics, both conceptually and in terms of technique.

In the 20th century, algebraic geometry split into several subareas.

- The mainstream of algebraic geometry is devoted to the study of the complex points of the algebraic varieties and more generally to the points with coordinates in an algebraically closed field.
- The study of the points of an algebraic variety with coordinates in the field of the rational numbers or in a number field became arithmetic geometry (or more classically Diophantine geometry), a subfield of algebraic number theory.
- The study of the real points of an algebraic variety is the subject of real algebraic geometry.

- A large part of singularity theory is devoted to the singularities of algebraic varieties.
- With the rise of the computers, a computational algebraic geometry area has emerged, which lies at the intersection of algebraic geometry and computer algebra. It consists essentially in developing algorithms and software for studying and finding the properties of explicitly given algebraic varieties.

Much of the development of the mainstream of algebraic geometry in the 20th century occurred within an abstract algebraic framework, with increasing emphasis being placed on "intrinsic" properties of algebraic varieties not dependent on any particular way of embedding the variety in an ambient coordinate space; this parallels developments in topology, differential and complex geometry. One key achievement of this abstract algebraic geometry is Grothendieck's scheme theory which allows one to use sheaf theory to study algebraic varieties in a way which is very similar to its use in the study of differential and analytic manifolds. This is obtained by extending the notion of point: In classical algebraic geometry, a point of an affine variety may be identified, through Hilbert's Nullstellensatz, with a maximal ideal of the coordinate ring, while the points of the corresponding affine scheme are all prime ideals of this ring. This means that a point of such a scheme may be either a usual point or a subvariety. This approach also enables a unification of the language and the tools of classical algebraic geometry, mainly concerned with complex points, and of algebraic number theory. Wiles's proof of the longstanding conjecture called Fermat's last theorem is an example of the power of this approach.

Basic Notions

Zeros of Simultaneous Polynomials

In classical algebraic geometry, the main objects of interest are the vanishing sets of collections of polynomials, meaning the set of all points that simultaneously satisfy one or more polynomial equations. For instance, the two-dimensional sphere of radius 1 in three-dimensional Euclidean space R^3 could be defined as the set of all points (x,y,z) with;

$$x^2 + y^2 + z^2 - 1 = 0.$$

A "slanted" circle in R^3 can be defined as the set of all points (x,y,z) which satisfy the two polynomial equations;

$$x^2 + y^2 + z^2 - 1 = 0,$$
$$x + y + z = 0.$$

Affine Varieties

First we start with a field k. In classical algebraic geometry, this field was always the complex numbers C, but many of the same results are true if we assume only that k is algebraically closed. We consider the affine space of dimension n over k, denoted $A^n(k)$ (or more simply A^n, when k is clear from the context). When one fixes a coordinate system, one may identify $A^n(k)$ with k^n. The purpose of not working with k^n is to emphasize that one "forgets" the vector space structure that k^n carries.

A function $f: A^n \to A^1$ is said to be *polynomial* (or *regular*) if it can be written as a polynomial, that is, if there is a polynomial p in $k[x_1,...,x_n]$ such that $f(M) = p(t_1,...,t_n)$ for every point M with coordinates $(t_1,...,t_n)$ in A^n. The property of a function to be polynomial (or regular) does not depend on the choice of a coordinate system in A^n.

When a coordinate system is chosen, the regular functions on the affine n-space may be identified with the ring of polynomial functions in n variables over k. Therefore, the set of the regular functions on A^n is a ring, which is denoted $k[A^n]$.

We say that a polynomial *vanishes* at a point if evaluating it at that point gives zero. Let S be a set of polynomials in $k[A^n]$. The *vanishing set of S* (or *vanishing locus* or *zero set*) is the set $V(S)$ of all points in A^n where every polynomial in S vanishes. Symbolically,

$$V(S) = \{(t_1,\dots,t_n) \mid \forall p \in S, p(t_1,\dots,t_n) = 0\}.$$

A subset of A^n which is $V(S)$, for some S, is called an *algebraic set*. The V stands for *variety* (a specific type of algebraic set to be defined below).

Given a subset U of A^n, can one recover the set of polynomials which generate it? If U is *any* subset of A^n, define $I(U)$ to be the set of all polynomials whose vanishing set contains U. The I stands for ideal: if two polynomials f and g both vanish on U, then $f+g$ vanishes on U, and if h is any polynomial, then hf vanishes on U, so $I(U)$ is always an ideal of the polynomial ring $k[A^n]$.

Two natural questions to ask are:

- Given a subset U of A^n, when is $U = V(I(U))$?
- Given a set S of polynomials, when is $S = I(V(S))$?

The answer to the first question is provided by introducing the Zariski topology, a topology on A^n whose closed sets are the algebraic sets, and which directly reflects the algebraic structure of $k[A^n]$. Then $U = V(I(U))$ if and only if U is an algebraic set or equivalently a Zariski-closed set. The answer to the second question is given by Hilbert's Nullstellensatz. In one of its forms, it says that $I(V(S))$ is the radical of the ideal generated by S. In more abstract language, there is a Galois connection, giving rise to two closure operators; they can be identified, and naturally play a basic role in the theory; the example is elaborated at Galois connection.

For various reasons we may not always want to work with the entire ideal corresponding to an algebraic set U. Hilbert's basis theorem implies that ideals in $k[A^n]$ are always finitely generated.

An algebraic set is called *irreducible* if it cannot be written as the union of two smaller algebraic sets. Any algebraic set is a finite union of irreducible algebraic sets and this decomposition is unique. Thus its elements are called the *irreducible components* of the algebraic set. An irreducible algebraic set is also called a *variety*. It turns out that an algebraic set is a variety if and only if it may be defined as the vanishing set of a prime ideal of the polynomial ring.

Some authors do not make a clear distinction between algebraic sets and varieties and use *irreducible variety* to make the distinction when needed.

Regular Functions

Just as continuous functions are the natural maps on topological spaces and smooth functions are the natural maps on differentiable manifolds, there is a natural class of functions on an algebraic set, called *regular functions* or *polynomial functions*. A regular function on an algebraic set V contained in A^n is the restriction to V of a regular function on A^n. For an algebraic set defined on the field of the complex numbers, the regular functions are smooth and even analytic.

It may seem unnaturally restrictive to require that a regular function always extend to the ambient space, but it is very similar to the situation in a normal topological space, where the Tietze extension theorem guarantees that a continuous function on a closed subset always extends to the ambient topological space.

Just as with the regular functions on affine space, the regular functions on V form a ring, which we denote by $k[V]$. This ring is called the *coordinate ring of V*.

Since regular functions on V come from regular functions on A^n, there is a relationship between the coordinate rings. Specifically, if a regular function on V is the restriction of two functions f and g in $k[A^n]$, then $f - g$ is a polynomial function which is null on V and thus belongs to $I(V)$. Thus $k[V]$ may be identified with $k[A^n]/I(V)$.

Morphism of Affine Varieties

Using regular functions from an affine variety to A^1, we can define regular maps from one affine variety to another. First we will define a regular map from a variety into affine space: Let V be a variety contained in A^n. Choose m regular functions on V, and call them $f_1, ..., f_m$. We define a *regular map f* from V to A^m by letting $f = (f_1, ..., f_m)$. In other words, each f_i determines one coordinate of the range of f.

If V' is a variety contained in A^m, we say that f is a *regular map* from V to V' if the range of f is contained in V'.

The definition of the regular maps apply also to algebraic sets. The regular maps are also called *morphisms*, as they make the collection of all affine algebraic sets into a category, where the objects are the affine algebraic sets and the morphisms are the regular maps. The affine varieties is a subcategory of the category of the algebraic sets.

Given a regular map g from V to V' and a regular function f of $k[V']$, then $f \circ g \in k[V]$. The map $f \rightarrow f \circ g$ is a ring homomorphism from $k[V']$ to $k[V]$. Conversely, every ring homomorphism from $k[V']$ to $k[V]$ defines a regular map from V to V'. This defines an equivalence of categories between the category of algebraic sets and the opposite category of the finitely generated reduced k-algebras. This equivalence is one of the starting points of scheme theory.

Rational Function and Birational Equivalence

In contrast to the preceding sections, this section concerns only varieties and not algebraic sets. On the other hand, the definitions extend naturally to projective varieties (next section), as an affine variety and its projective completion have the same field of functions.

If V is an affine variety, its coordinate ring is an integral domain and has thus a field of fractions which is denoted $k(V)$ and called the *field of the rational functions* on V or, shortly, the *function field* of V. Its elements are the restrictions to V of the rational functions over the affine space containing V. The domain of a rational function f is not V but the complement of the subvariety (a hypersurface) where the denominator of f vanishes.

As with regular maps, one may define a *rational map* from a variety V to a variety V'. As with the regular maps, the rational maps from V to V' may be identified to the field homomorphisms from $k(V')$ to $k(V)$.

Two affine varieties are *birationally equivalent* if there are two rational functions between them which are inverse one to the other in the regions where both are defined. Equivalently, they are birationally equivalent if their function fields are isomorphic.

An affine variety is a *rational variety* if it is birationally equivalent to an affine space. This means that the variety admits a rational parameterization. For example, the circle of equation $x^2 + y^2 - 1 = 0$ is a rational curve, as it has the parameterization:

$$x = \frac{2t}{1+t^2}$$

$$y = \frac{1-t^2}{1+t^2},$$

which may also be viewed as a rational map from the line to the circle.

The problem of resolution of singularities is to know if every algebraic variety is birationally equivalent to a variety whose projective completion is nonsingular. It was solved in the affirmative in characteristic 0 by Heisuke Hironaka in 1964 and is yet unsolved in finite characteristic.

Projective Variety

Just as the formulas for the roots of second, third, and fourth degree polynomials suggest extending real numbers to the more algebraically complete setting of the complex numbers, many properties of algebraic varieties suggest extending affine space to a more geometrically complete projective space. Whereas the complex numbers are obtained by adding the number i, a root of the polynomial $x^2 + 1$, projective space is obtained by adding in appropriate points "at infinity", points where parallel lines may meet.

To see how this might come about, consider the variety $V(y - x^2)$. If we draw it, we get a parabola. As x goes to positive infinity, the slope of the line from the origin to the point (x, x^2) also goes to positive infinity. As x goes to negative infinity, the slope of the same line goes to negative infinity.

Compare this to the variety $V(y - x^3)$. This is a cubic curve. As x goes to positive infinity, the slope of the line from the origin to the point (x, x^3) goes to positive infinity just as before. But unlike before, as x goes to negative infinity, the slope of the same line goes to positive infinity as well; the exact opposite of the parabola. So the behavior "at infinity" of $V(y - x^3)$ is different from the behavior "at infinity" of $V(y - x^2)$.

The consideration of the *projective completion* of the two curves, which is their prolongation "at infinity" in the projective plane, allows us to quantify this difference: the point at infinity of the parabola is a regular point, whose tangent is the line at infinity, while the point at infinity of the cubic curve is a cusp. Also, both curves are rational, as they are parameterized by x, and the Riemann-Roch theorem implies that the cubic curve must have a singularity, which must be at infinity, as all its points in the affine space are regular.

Thus many of the properties of algebraic varieties, including birational equivalence and all the topological properties, depend on the behavior "at infinity" and so it is natural to study the varieties in projective space. Furthermore, the introduction of projective techniques made many theorems in algebraic geometry simpler and sharper: For example, Bézout's theorem on the number of intersection points between two varieties can be stated in its sharpest form only in projective space. For these reasons, projective space plays a fundamental role in algebraic geometry.

Nowadays, the *projective space* P^n of dimension n is usually defined as the set of the lines passing through a point, considered as the origin, in the affine space of dimension $n + 1$, or equivalently to the set of the vector lines in a vector space of dimension $n + 1$. When a coordinate system has been chosen in the space of dimension $n + 1$, all the points of a line have the same set of coordinates, up to the multiplication by an element of k. This defines the homogeneous coordinates of a point of P^n as a sequence of $n + 1$ elements of the base field k, defined up to the multiplication by a nonzero element of k (the same for the whole sequence).

A polynomial in $n + 1$ variables vanishes at all points of a line passing through the origin if and only if it is homogeneous. In this case, one says that the polynomial *vanishes* at the corresponding point of P^n. This allows us to define a *projective algebraic set* in P^n as the set $V(f_1, ..., f_k)$, where a finite set of homogeneous polynomials $\{f_1, ..., f_k\}$ vanishes. Like for affine algebraic sets, there is a bijection between the projective algebraic sets and the reduced homogeneous ideals which define them. The *projective varieties* are the projective algebraic sets whose defining ideal is prime. In other words, a projective variety is a projective algebraic set, whose homogeneous coordinate ring is an integral domain, the *projective coordinates ring* being defined as the quotient of the graded ring or the polynomials in $n + 1$ variables by the homogeneous (reduced) ideal defining the variety. Every projective algebraic set may be uniquely decomposed into a finite union of projective varieties.

The only regular functions which may be defined properly on a projective variety are the constant functions. Thus this notion is not used in projective situations. On the other hand, the *field of the rational functions* or *function field* is a useful notion, which, similarly to the affine case, is defined as the set of the quotients of two homogeneous elements of the same degree in the homogeneous coordinate ring.

Real Algebraic Geometry

Real algebraic geometry is the study of the real points of algebraic geometry.

The fact that the field of the real numbers is an ordered field cannot be ignored in such a study. For example, the curve of equation $x^2 + y^2 - a = 0$ is a circle if $a > 0$, but does not have any real point if $a < 0$. It follows that real algebraic geometry is not only the study of the real algebraic varieties, but has been generalized to the study of the *semi-algebraic sets*, which are the solutions of systems of polynomial equations and polynomial inequalities. For example, a branch of the hyperbola of equation $xy-1=0$ is not an algebraic variety, but is a semi-algebraic set defined by $xy-1=0$ and $x>0$ or by $xy-1=0$ and $x+y>0$.

One of the challenging problems of real algebraic geometry is the unsolved Hilbert's sixteenth problem: Decide which respective positions are possible for the ovals of a nonsingular plane curve of degree 8.

Computational Algebraic Geometry

One may date the origin of computational algebraic geometry to meeting EUROSAM'79 (International Symposium on Symbolic and Algebraic Manipulation) held at Marseille, France in June 1979. At this meeting,

- Dennis S. Arnon showed that George E. Collins's Cylindrical algebraic decomposition (CAD) allows the computation of the topology of semi-algebraic sets,
- Bruno Buchberger presented the Gröbner bases and his algorithm to compute them,
- Daniel Lazard presented a new algorithm for solving systems of homogeneous polynomial equations with a computational complexity which is essentially polynomial in the expected number of solutions and thus simply exponential in the number of the unknowns. This algorithm is strongly related with Macaulay's multivariate resultant.

Since then, most results in this area are related to one or several of these items either by using or improving one of these algorithms, or by finding algorithms whose complexity is simply exponential in the number of the variables.

A body of mathematical theory complementary to symbolic methods called numerical algebraic geometry has been developed over the last several decades. The main computational method is homotopy continuation. This supports, for example, a model of floating point computation for solving problems of algebraic geometry.

Gröbner Basis

A Gröbner basis is a system of generators of a polynomial ideal whose computation allows the deduction of many properties of the affine algebraic variety defined by the ideal.

Given an ideal *I* defining an algebraic set *V*:

- *V* is empty (over an algebraically closed extension of the basis field), if and only if the Gröbner basis for any monomial ordering is reduced to {1}.
- By means of the Hilbert series one may compute the dimension and the degree of *V* from any Gröbner basis of *I* for a monomial ordering refining the total degree.
- If the dimension of *V* is 0, one may compute the points (finite in number) of *V* from any Gröbner basis of *I*.
- A Gröbner basis computation allows one to remove from *V* all irreducible components which are contained in a given hyper-surface.

- A Gröbner basis computation allows one to compute the Zariski closure of the image of V by the projection on the k first coordinates, and the subset of the image where the projection is not proper.
- More generally Gröbner basis computations allow one to compute the Zariski closure of the image and the critical points of a rational function of V into another affine variety.

Gröbner basis computations do not allow one to compute directly the primary decomposition of I nor the prime ideals defining the irreducible components of V, but most algorithms for this involve Gröbner basis computation. The algorithms which are not based on Gröbner bases use regular chains but may need Gröbner bases in some exceptional situations.

Gröbner bases are deemed to be difficult to compute. In fact they may contain, in the worst case, polynomials whose degree is doubly exponential in the number of variables and a number of polynomials which is also doubly exponential. However, this is only a worst case complexity, and the complexity bound of Lazard's algorithm of 1979 may frequently apply. Faugère F5 algorithm realizes this complexity, as it may be viewed as an improvement of Lazard's 1979 algorithm. It follows that the best implementations allow one to compute almost routinely with algebraic sets of degree more than 100. This means that, presently, the difficulty of computing a Gröbner basis is strongly related to the intrinsic difficulty of the problem.

Cylindrical Algebraic Decomposition (CAD)

CAD is an algorithm which was introduced in 1973 by G. Collins to implement with an acceptable complexity the Tarski–Seidenberg theorem on quantifier elimination over the real numbers.

This theorem concerns the formulas of the first-order logic whose atomic formulas are polynomial equalities or inequalities between polynomials with real coefficients. These formulas are thus the formulas which may be constructed from the atomic formulas by the logical operators *and* ($\wedge$), *or* ($\vee$), *not* ($\neg$), *for all* ($\forall$) and *exists* ($\exists$). Tarski's theorem asserts that, from such a formula, one may compute an equivalent formula without quantifier ($\forall$, $\exists$).

The complexity of CAD is doubly exponential in the number of variables. This means that CAD allows, in theory, to solve every problem of real algebraic geometry which may be expressed by such a formula, that is almost every problem concerning explicitly given varieties and semi-algebraic sets.

While Gröbner basis computation has doubly exponential complexity only in rare cases, CAD has almost always this high complexity. This implies that, unless if most polynomials appearing in the input are linear, it may not solve problems with more than four variables.

Since 1973, most of the research on this subject is devoted either to improve CAD or to find alternate algorithms in special cases of general interest.

As an example of the state of art, there are efficient algorithms to find at least a point in every connected component of a semi-algebraic set, and thus to test if a semi-algebraic set is empty. On the other hand, CAD is yet, in practice, the best algorithm to count the number of connected components.

Asymptotic Complexity vs. Practical Efficiency

The basic general algorithms of computational geometry have a double exponential worst case complexity. More precisely, if d is the maximal degree of the input polynomials and n the number of variables, their complexity is at most $d^{2^{cn}}$ for some constant c, and, for some inputs, the complexity is at least $d^{2^{c'n}}$ for another constant c'.

During the last 20 years of 20th century, various algorithms have been introduced to solve specific subproblems with a better complexity. Most of these algorithms have a complexity $d^{O(n^2)}$.

Among these algorithms which solve a sub problem of the problems solved by Gröbner bases, one may cite *testing if an affine variety is empty* and *solving nonhomogeneous polynomial systems which have a finite number of solutions.* Such algorithms are rarely implemented because, on most entries Faugère's F4 and F5 algorithms have a better practical efficiency and probably a similar or better complexity (*probably* because the evaluation of the complexity of Gröbner basis algorithms on a particular class of entries is a difficult task which has been done only in a few special cases).

The main algorithms of real algebraic geometry which solve a problem solved by CAD are related to the topology of semi-algebraic sets. One may cite *counting the number of connected components, testing if two points are in the same components* or *computing a Whitney stratification of a real algebraic set*. They have a complexity of $d^{O(n^2)}$, but the constant involved by O notation is so high that using them to solve any nontrivial problem effectively solved by CAD, is impossible even if one could use all the existing computing power in the world. Therefore, these algorithms have never been implemented and this is an active research area to search for algorithms with have together a good asymptotic complexity and a good practical efficiency.

Abstract Modern Viewpoint

The modern approaches to algebraic geometry redefine and effectively extend the range of basic objects in various levels of generality to schemes, formal schemes, ind-schemes, algebraic spaces, algebraic stacks and so on. The need for this arises already from the useful ideas within theory of varieties, e.g. the formal functions of Zariski can be accommodated by introducing nilpotent elements in structure rings; considering spaces of loops and arcs, constructing quotients by group actions and developing formal grounds for natural intersection theory and deformation theory lead to some of the further extensions.

Most remarkably, in late 1950s, algebraic varieties were subsumed into Alexander Grothendieck's concept of a scheme. Their local objects are affine schemes or prime spectra which are locally ringed spaces which form a category which is antiequivalent to the category of commutative unital rings, extending the duality between the category of affine algebraic varieties over a field k, and the category of finitely generated reduced k-algebras. The gluing is along Zariski topology; one can glue within the category of locally ringed spaces, but also, using the Yoneda embedding, within the more abstract category of presheaves of sets over the category of affine schemes. The Zariski topology in the set theoretic sense is then replaced by a Grothendieck topology. Grothendieck introduced Grothendieck topologies having in mind more exotic but geometrically finer and more sensitive examples than the crude Zariski topology, namely the étale topology, and the two flat Grothendieck topologies: fppf and fpqc; nowadays some other examples became prominent including Nisnevich topology. Sheaves can be furthermore generalized to stacks in the sense of Grothendieck, usually with some additional representability conditions leading to Artin stacks and, even finer, Deligne-Mumford stacks, both often called algebraic stacks.

Sometimes other algebraic sites replace the category of affine schemes. For example, Nikolai Durov has introduced commutative algebraic monads as a generalization of local objects in a generalized algebraic geometry. Versions of a tropical geometry, of an absolute geometry over a field of one element and an algebraic analogue of Arakelov's geometry were realized in this setup.

Another formal generalization is possible to universal algebraic geometry in which every variety of algebras has its own algebraic geometry. The term *variety of algebras* should not be confused with *algebraic variety*.

The language of schemes, stacks and generalizations has proved to be a valuable way of dealing with geometric concepts and became cornerstones of modern algebraic geometry.

Algebraic stacks can be further generalized and for many practical questions like deformation theory and intersection theory, this is often the most natural approach. One can extend the Grothendieck site of affine schemes to a higher categorical site of derived affine schemes, by replacing the commutative rings with an infinity category of differential graded commutative algebras, or of simplicial commutative rings or a similar category with an appropriate variant of a Grothendieck topology. One can also replace presheaves of sets by presheaves of simplicial sets (or of infinity groupoids). Then, in presence of an appropriate homotopic machinery one can develop a notion of derived stack as such a presheaf on the infinity category of derived affine schemes, which is satisfying certain infinite categorical version of a sheaf axiom (and to be algebraic, inductively a sequence of representability conditions). Quillen model categories, Segal categories and quasi-categories are some of the most often used tools to formalize this yielding the *derived algebraic geometry*, introduced by the school of Carlos Simpson, including Andre Hirschowitz, Bertrand Toën, Gabrielle Vezzosi, Michel Vaquié and others; and developed further by Jacob Lurie, Bertrand Toën, and Gabrielle Vezzosi. Another (noncommutative) version of derived algebraic geometry, using A-infinity categories has been developed from early 1990s by Maxim Kontsevich and followers.

Analytic Geometry

An analytic variety is defined locally as the set of common solutions of several equations involving analytic functions. It is analogous to the included concept of real or complex algebraic variety. Any complex manifold is an analytic variety. Since analytic varieties may have singular points, not all analytic varieties are manifolds.

Modern analytic geometry is essentially equivalent to real and complex algebraic geometry, as has been shown by Jean-Pierre Serre in his paper *GAGA*, the name of which is French for *Algebraic geometry and analytic geometry*. Nevertheless, the two fields remain distinct, as the methods of proof are quite different and algebraic geometry includes also geometry in finite characteristic.

Applications

Algebraic geometry now finds applications in statistics, control theory, robotics, error-correcting codes, phylogenetics and geometric modelling. There are also connections to string theory, game theory, graph matchings, solitons and integer programming.

Algebraic Variety

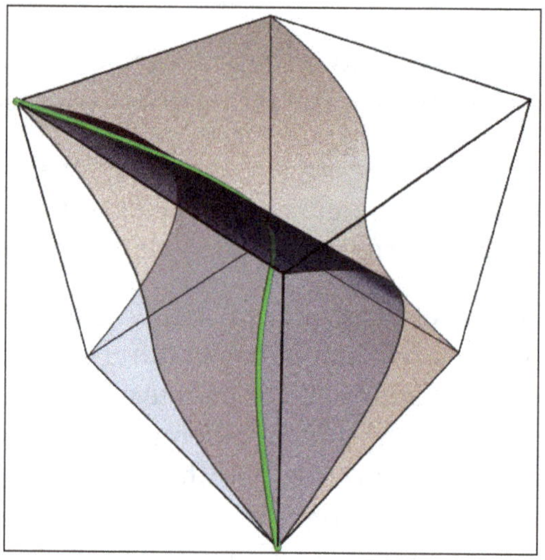

The twisted cubic is a projective algebraic variety.

Algebraic varieties are the central objects of study in algebraic geometry. Classically, an algebraic variety is defined as the set of solutions of a system of polynomial equations over the real or complex numbers. Modern definitions generalize this concept in several different ways, while attempting to preserve the geometric intuition behind the original definition.

Conventions regarding the definition of an algebraic variety differ slightly. For example, some definitions require an algebraic variety to be irreducible, which means that it is not the union of two smaller sets that are closed in the Zariski topology. Under this definition, non-irreducible algebraic varieties are called algebraic sets. Other conventions do not require irreducibility.

The concept of an algebraic variety is similar to that of an analytic manifold. An important difference is that an algebraic variety may have singular points, while a manifold cannot.

The fundamental theorem of algebra establishes a link between algebra and geometry by showing that a monic polynomial (an algebraic object) in one variable with complex number coefficients is determined by the set of its roots (a geometric object) in the complex plane. Generalizing this result, Hilbert's Nullstellensatz provides a fundamental correspondence between ideals of polynomial rings and algebraic sets. Using the Nullstellensatz and related results, mathematicians have established a strong correspondence between questions on algebraic sets and questions of ring theory. This correspondence is a defining feature of algebraic geometry.

Overview and Definitions

An *affine variety* over an algebraically closed field is conceptually the easiest type of variety to define, which will be done in this section. Next, one can define projective and quasi-projective varieties in a similar way. The most general definition of a variety is obtained by patching together smaller quasi-projective varieties. It is not obvious that one can construct genuinely new examples of varieties in this way, but Nagata gave an example of such a new variety in the 1950s.

Affine Varieties

For an algebraically closed field k and a natural number n, let A^n be affine n-space over k. The polynomials f in the ring $k[x_1, ..., x_n]$ can be viewed as k-valued functions on A^n by evaluating f at the points in A^n, i.e. by choosing values in k for each x_i. For each set S of polynomials in $k[x_1, ..., x_n]$, define the zero-locus $Z(S)$ to be the set of points in A^n on which the functions in S simultaneously vanish, that is to say;

$$Z(S) = \left\{x \in \mathbf{A}^n \mid f(x) = 0 \text{ for all } f \in S\right\}.$$

A subset V of A^n is called an affine algebraic set if $V = Z(S)$ for some S. A nonempty affine algebraic set V is called irreducible if it cannot be written as the union of two proper algebraic subsets. An irreducible affine algebraic set is also called an affine variety. (Many authors use the phrase *affine variety* to refer to any affine algebraic set, irreducible or not);

Affine varieties can be given a natural topology by declaring the closed sets to be precisely the affine algebraic sets. This topology is called the Zariski topology.

Given a subset V of A^n, we define $I(V)$ to be the ideal of all polynomial functions vanishing on V:

$$I(V) = \left\{f \in k[x_1, \ldots, x_n] \mid f(x) = 0 \text{ for all } x \in V\right\}.$$

For any affine algebraic set V, the coordinate ring or structure ring of V is the quotient of the polynomial ring by this ideal.

Projective Varieties and Quasi-projective Varieties

Let k be an algebraically closed field and let P^n be the projective n-space over k. Let f in $k[x_0, ..., x_n]$ be a homogeneous polynomial of degree d. It is not well-defined to evaluate f on points in P^n in homogeneous coordinates. However, because f is homogeneous, meaning that $f(\lambda x_0, ..., \lambda x_n) = \lambda^d f(x_0, ..., x_n)$, it *does* make sense to ask whether f vanishes at a point $[x_0 : ... : x_n]$. For each set S of homogeneous polynomials, define the zero-locus of S to be the set of points in P^n on which the functions in S vanish:

$$Z(S) = \{x \in \mathbf{P}^n \mid f(x) = 0 \text{ for all } f \in S\}.$$

A subset V of P^n is called a projective algebraic set if $V = Z(S)$ for some S. An irreducible projective algebraic set is called a projective variety.

Projective varieties are also equipped with the Zariski topology by declaring all algebraic sets to be closed.

Given a subset V of P^n, let $I(V)$ be the ideal generated by all homogeneous polynomials vanishing on V. For any projective algebraic set V, the coordinate ring of V is the quotient of the polynomial ring by this ideal.

A quasi-projective variety is a Zariski open subset of a projective variety. Notice that every affine variety is quasi-projective. Notice also that the complement of an algebraic set in an affine variety is a quasi-projective variety; in the context of affine varieties, such a quasi-projective variety is usually not called a variety but a constructible set.

Abstract Varieties

In classical algebraic geometry, all varieties were by definition quasiprojective varieties, meaning that they were open subvarieties of closed subvarieties of projective space. For example, in Chapter 1 of Hartshorne a *variety* over an algebraically closed field is defined to be a quasi-projective variety, but from Chapter 2 onwards, the term variety (also called an abstract variety) refers to a more general object, which locally is a quasi-projective variety, but when viewed as a whole is not necessarily quasi-projective; i.e. it might not have an embedding into projective space. So classically the definition of an algebraic variety required an embedding into projective space, and this embedding was used to define the topology on the variety and the regular functions on the variety. The disadvantage of such a definition is that not all varieties come with natural embeddings into projective space. For example, under this definition, the product $P^1 \times P^1$ is not a variety until it is embedded into the projective space; this is usually done by the Segre embedding. However, any variety that admits one embedding into projective space admits many others by composing the embedding with the Veronese embedding. Consequently, many notions that should be intrinsic, such as the concept of a regular function, are not obviously so.

The earliest successful attempt to define an algebraic variety abstractly, without an embedding, was made by André Weil. In his *Foundations of Algebraic Geometry*, Weil defined an abstract algebraic variety using valuations. Claude Chevalley made a definition of a scheme, which served a similar purpose, but was more general. However, Alexander Grothendieck's definition of a scheme is more general still and has received the most widespread acceptance. In Grothendieck's language, an abstract algebraic variety is usually defined to be an integral, separated scheme of finite type over an algebraically closed field, although some authors drop the irreducibility or the reducedness or the separateness condition or allow the underlying field to be not algebraically closed. Classical algebraic varieties are the quasiprojective integral separated finite type schemes over an algebraically closed field.

Existence of Non-quasiprojective Abstract Algebraic Varieties

One of the earliest examples of a non-quasiprojective algebraic variety were given by Nagata. Nagata's example was not complete (the analog of compactness), but soon afterwards he found an algebraic surface that was complete and non-projective. Since then other examples have been found.

Examples

Subvariety

A subvariety is a subset of a variety that is itself a variety (with respect to the structure induced from the ambient variety). For example, every open subset of a variety is a variety.

Hilbert's Nullstellensatz says that closed subvarieties of an affine or projective variety are in one-to-one correspondence with the prime ideals or homogeneous prime ideals of the coordinate ring of the variety.

Affine Variety

Example 1

Let k = C, and A^2 be the two-dimensional affine space over C. Polynomials in the ring C[x, y] can be viewed as complex valued functions on A^2 by evaluating at the points in A^2. Let subset S of C[x, y] contain a single element f (x, y):

$$f(x,y)=x+y-1.$$

The zero-locus of f (x, y) is the set of points in A^2 on which this function vanishes: it is the set of all pairs of complex numbers (x, y) such that $y = 1 - x$. This is called a line in the affine plane. (In the classical topology coming from the topology on the complex numbers, a complex line is a real manifold of dimension two.) This is the set $Z(f)$:

$$Z(f)=\{(x,1-x)\in \mathbf{C}^2\}.$$

Thus the subset $V = Z(f)$ of A^2 is an algebraic set. The set V is not empty. It is irreducible, as it cannot be written as the union of two proper algebraic subsets. Thus it is an affine algebraic variety.

Example 2

Let k = C, and A^2 be the two-dimensional affine space over C. Polynomials in the ring C[x, y] can be viewed as complex valued functions on A^2 by evaluating at the points in A^2. Let subset S of C[x, y] contain a single element $g(x, y)$:

$$g(x,y)=x^2+y^2-1.$$

The zero-locus of $g(x, y)$ is the set of points in A^2 on which this function vanishes, that is the set of points (x,y) such that $x^2 + y^2 = 1$. As $g(x, y)$ is an absolutely irreducible polynomial, this is an algebraic variety. The set of its real points (that is the points for which x and y are real numbers), is known as the unit circle; this name is also often given to the whole variety.

Example 3

The following example is neither a hypersurface, nor a linear space, nor a single point. Let A^3 be the three-dimensional affine space over C. The set of points (x, x^2, x^3) for x in C is an algebraic variety, and more precisely an algebraic curve that is not contained in any plane. It is the twisted cubic shown in the above figure. It may be defined by the equations;

$$y-x^2=0$$
$$z-x^3=0$$

The irreducibility of this algebraic set needs a proof. One approach in this case is to check that the projection $(x, y, z) \rightarrow (x, y)$ is injective on the set of the solutions and that its image is an irreducible plane curve.

For more difficult examples, a similar proof may always be given, but may imply a difficult computation: first a Gröbner basis computation to compute the dimension, followed by a random linear change of variables (not always needed); then a Gröbner basis computation for another monomial ordering to compute the projection and to prove that it is generically injective and that its image is a hypersurface, and finally a polynomial factorization to prove the irreducibility of the image.

Projective Variety

A projective variety is a closed subvariety of a projective space. That is, it is the zero locus of a set of homogeneous polynomials that generate a prime ideal.

Example 1

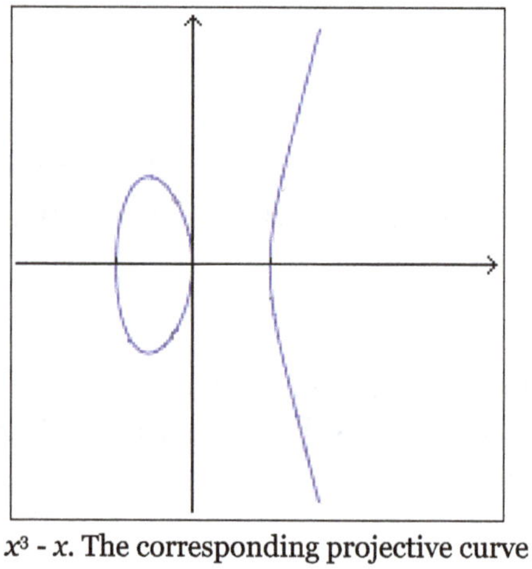

The affine plane curve $y^2 = x^3 - x$. The corresponding projective curve is called an elliptic curve.

A plane projective curve is the zero locus of an irreducible homogeneous polynomial in three indeterminates. The projective line P^1 is an example of a projective curve; it can be viewed as the curve in the projective plane $P^2 = \{[x,y,z]\}$ defined by $x = 0$. For another example, first consider the affine cubic curve;

$$y^2 = x^3 - x.$$

in the 2-dimensional affine space (over a field of characteristic not two). It has the associated cubic homogeneous polynomial equation:

$$y^2 z = x^3 - xz^2,$$

which defines a curve in P^2 called an elliptic curve. The curve has genus one (genus formula); in particular, it is not isomorphic to the projective line P^1, which has genus zero. Using genus to distinguish curves is very basic: in fact, the genus is the first invariant one uses to classify curves.

Example 2

Let V be a finite-dimensional vector space. The Grassmannian variety $G_n(V)$ is the set of all n-dimensional subspaces of V. It is a projective variety: it is embedded into a projective space via the Plücker embedding:

$$G_n(V) \hookrightarrow \mathbf{P}(\wedge^n V), \langle b_1, \ldots, b_n \rangle \mapsto [b_1 \wedge \cdots \wedge b_n]$$

where b_i are any set of linearly independent vectors in V, $\wedge^n V$ is the n-th exterior power of V and the bracket $[w]$ means the line spanned by the nonzero vector w.

The Grassmannian variety comes with a natural vector bundle (or locally free sheaf in other terminology) called the tautological bundle, which is important in the study of characteristic classes such as Chern classes.

Non-affine and Non-projective Example

An algebraic variety can be neither affine nor projective. To give an example, let $X = P^1 \times A^1$ and p: $X \to A^1$ the projection. It is an algebraic variety since it is a product of varieties. It is not affine since P^1 is a closed subvariety of X (as the zero locus of p), but an affine variety cannot contain a projective variety of positive dimension as a closed subvariety. It is not projective either, since there is a nonconstant regular function on X; namely, p.

Another example of a non-affine non-projective variety is $X = A^2 - (0, 0)$ (cf. morphism of varieties#Examples.)

Basic Results

- An affine algebraic set V is a variety if and only if $I(V)$ is a prime ideal; equivalently, V is a variety if and only if its coordinate ring is an integral domain.
- Every nonempty affine algebraic set may be written uniquely as a finite union of algebraic varieties (where none of the varieties in the decomposition is a subvariety of any other).
- The dimension of a variety may be defined in various equivalent ways.

Isomorphism of Algebraic Varieties

Let V_1, V_2 be algebraic varieties. We say V_1 and V_2 are isomorphic, and write $V_1 \cong V_2$, if there are regular maps $\varphi : V_1 \to V_2$ and $\psi : V_2 \to V_1$ such that the compositions $\psi \circ \varphi$ and $\varphi \circ \psi$ are the identity maps on V_1 and V_2 respectively.

Discussion and Generalizations

The basic definitions and facts above enable one to do classical algebraic geometry. To be able to do more — for example, to deal with varieties over fields that are not algebraically closed — some foundational changes are required. The modern notion of a variety is considerably more abstract than the one above, though equivalent in the case of varieties over algebraically closed fields. An *abstract algebraic variety* is a particular kind of scheme; the generalization to schemes on the geometric side enables an extension of the correspondence described above to a wider class of rings. A scheme is a locally ringed space such that every point has a neighbourhood that, as a locally ringed space, is isomorphic to a spectrum of a ring. Basically, a variety over k is a scheme whose structure sheaf is a sheaf of k-algebras with the property that the rings R that occur above are all integral domains and are all finitely generated k-algebras, that is to say, they are quotients of polynomial algebras by prime ideals.

This definition works over any field k. It allows you to glue affine varieties (along common open sets) without worrying whether the resulting object can be put into some projective space. This also leads to difficulties since one can introduce somewhat pathological objects, e.g. an affine line with zero doubled. Such objects are usually not considered varieties, and are eliminated by requiring the schemes underlying a variety to be *separated.* (Strictly speaking, there is also a third condition, namely, that one needs only finitely many affine patches in the definition above.)

Some modern researchers also remove the restriction on a variety having integral domain affine charts, and when speaking of a variety only require that the affine charts have trivial nilradical.

A complete variety is a variety such that any map from an open subset of a nonsingular curve into it can be extended uniquely to the whole curve. Every projective variety is complete, but not vice versa.

These varieties have been called "varieties in the sense of Serre", since Serre's foundational paper FAC on sheaf cohomology was written for them. They remain typical objects to start studying in algebraic geometry, even if more general objects are also used in an auxiliary way.

One way that leads to generalizations is to allow reducible algebraic sets (and fields k that aren't algebraically closed), so the rings R may not be integral domains. A more significant modification is to allow nilpotents in the sheaf of rings, that is, rings which are not reduced. This is one of several generalizations of classical algebraic geometry that are built into Grothendieck's theory of schemes.

Allowing nilpotent elements in rings is related to keeping track of "multiplicities" in algebraic geometry. For example, the closed subscheme of the affine line defined by $x^2 = 0$ is different from the subscheme defined by $x = 0$ (the origin). More generally, the fiber of a morphism of schemes $X \to Y$ at a point of Y may be non-reduced, even if X and Y are reduced. Geometrically, this says that fibers of good mappings may have nontrivial "infinitesimal" structure.

There are further generalizations called algebraic spaces and stacks.

Algebraic Manifolds

An algebraic manifold is an algebraic variety that is also an m-dimensional manifold, and hence every sufficiently small local patch is isomorphic to k^m. Equivalently, the variety is smooth (free from singular points). When k is the real numbers, R, algebraic manifolds are called Nash manifolds. Algebraic manifolds can be defined as the zero set of a finite collection of analytic algebraic functions. Projective algebraic manifolds are an equivalent definition for projective varieties. The Riemann sphere is one example.

Affine Variety

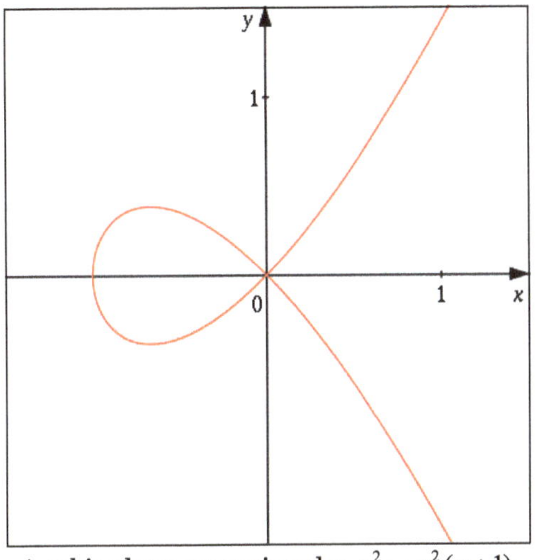

A cubic plane curve given by $y^2 = x^2(x+1)$.

In algebraic geometry, an affine variety over an algebraically closed field *k* is the zero-locus in the affine *n*-space k^n of some finite family of polynomials of *n* variables with coefficients in *k* that generate a prime ideal. If the condition of generating a prime ideal is removed, such a set is called an (affine) algebraic set. A Zariski open subvariety of an affine variety is called a quasi-affine variety.

If *X* is an affine variety defined by a prime ideal *I*, then the quotient ring:

$$k[x_1,\ldots,x_n]/I$$

is called the coordinate ring of *X*. This ring is precisely the set of all regular functions on *X*; in other words, it is the space of global sections of the structure sheaf of *X*. A theorem of Serre gives a cohomological characterization of an affine variety; it says an algebraic variety is affine if and only if:

$$H^i(X,F)=0$$

for any $i>0$ and any quasi-coherent sheaf *F* on *X*. (cf. Cartan's theorem B.) This makes the cohomological study of an affine variety non-existent, in a sharp contrast to the projective case in which cohomology groups of line bundles are of central interest.

An affine variety plays a role of a local chart for algebraic varieties; that is to say, general algebraic varieties such as projective varieties are obtained by gluing affine varieties. Linear structures that are attached to varieties are also (trivially) affine varieties; e.g., tangent spaces, fibers of algebraic vector bundles.

An affine variety is, up to an equivalence of categories a special case of an affine scheme, which is precisely the spectrum of a ring. In complex geometry, an affine variety is an analog of a Stein manifold.

Introduction

The most concrete point of view to describe an affine algebraic variety is that it is the set of solutions in an algebraically closed field k of a system of polynomial equations with coefficients in k. More precisely, if $f_1,\ldots,f_k$ are polynomials with coefficients in k, they define an affine variety (or affine algebraic set):

$$V(f_1,\ldots,f_k)=\left\{(a_1,\ldots,a_n)\in k^n \mid f_1(a_1,\ldots,a_n)=\ldots=f_k(a_1,\ldots,a_n)=0\right\}.$$

By Hilbert's Nullstellensatz, the points of the variety are in one-to-one correspondence with the maximal ideals of its *coordinate ring*, the k-algebra $R=k[x_1,\ldots,x_n]/\langle f_1,\ldots,f_k\rangle$, through the map $(a_1,\ldots,a_n)\mapsto\langle\overline{x_1-a_1},\ldots,\overline{x_n-a_n}\rangle$, where $\overline{x_i-a_i}$ denotes the image in the quotient algebra R of the polynomial x_i-a_i. In scheme theory, this correspondence has been extended to prime ideals to define the affine scheme $\operatorname{Spec}(R)$, which may be identified to the variety, through an equivalence of categories.

The elements of the coordinate ring R are also called the *regular functions* or the *polynomial functions* on the variety. They form the *ring of the regular functions* on the variety, or, simply, the *ring of the variety*. In fact an element $\overline{f}\in R$ is the image of a polynomial $f\in k[x_1,\ldots,x_n]$, which defines a function from k^n into k; The restriction of f to the variety does not depend on the choice of f among the polynomials mapped on $\overline{f}$ by the quotient.

The dimension of a variety is an integer associated to every variety, and even to every algebraic set, whose importance relies on the large number of its equivalent definitions.

Structure Sheaf

Equipped with the structure sheaf described below, an affine variety is a locally ringed space.

Given an affine variety X with coordinate ring A, we define the sheaf of k-algebras $\mathcal{O}_X$ by letting $\mathcal{O}_X(U)=\Gamma(U,\mathcal{O}_X)$ be the ring of regular functions on U.

We let $D(f) = \{ x \mid f(x) \neq 0 \}$ for each f in A. They form a base for the topology of X and so $\mathcal{O}_X$ is determined by its values on the open sets $D(f)$.

The key fact, which relies on Hilbert nullstellensatz in the essential way, is the following:

Claim — $\Gamma(D(f),\mathcal{O}_X)=A[f^{-1}]$ for any f in A.

Proof: The inclusion $\supset$ is clear. For the opposite, let g be in the left-hand side and $J=\{h\in A \mid hg\in A\}$, which is an ideal. If x is in $D(f)$, then, since g is regular near x, there is some open affine neighborhood $D(h)$ of x such that $g\in k[D(h)]=A[h^{-1}]$; that is, $h^m g$ is in A and thus x is not in $V(J)$. In other words, $V(J)\subset\{x \mid f(x)=0\}$ and thus the Hilbert nullstellensatz implies f is in the radical of J; i.e., $f^n g\in A$.

The claim, first of all, implies that X is a "locally ringed" space since:

$$\mathcal{O}_{X,x}=\varinjlim_{f(x)\neq 0} A[f^{-1}]=A_{\mathfrak{m}_x}$$

where $\mathfrak{m}_x=\{f\in A \mid f(x)=0\}$. Secondly, the claim implies that $\mathcal{O}_X$ is a sheaf; indeed, it says if a function is regular (pointwise) on $D(f)$, then it must be in the coordinate ring of $D(f)$; that is, "regular-ness" can be patched together.

Hence, $(X,\mathcal{O}_X)$ is a locally ringed space.

Examples

- The complement of a hypersurface of an affine variety X; i.e., $X - \{f = 0\}$ for some regular function f on X is affine; its coordinate ring is the localization $k[X][f^{-1}]$. In particular, $A^1 - 0$ (the affine line with the origin removed) is affine.
- Every closed subvariety of the affine space $\mathbf{A}^n$ of codimension one is defined by a prime ideal of the polynomial ring of height one, which is principal; thus, they are hypersurfaces (i.e., defined by a single polynomial.)
- $C^2 - 0$ is an open subset of the affine variety that is not affine; cf. Hartogs' extension theorem
- The normalization of an irreducible affine variety is affine; the coordinate ring of the normalization is the integral closure of the coordinate ring of the variety. (It turns out the normalization of a projective variety is a projective variety.)

Tangent Space

Tangent spaces may be defined just as in calculus. Let $X = \operatorname{spec} A, A = k[x_1,\dots,x_n]/(f_1,\dots,f_r)$ be the affine variety. Then the affine subvariety of k^n defined by the linear equations:

$$\sum_{i=1}^{n} \frac{\partial f_j}{\partial x_i}(a_1,\dots,a_n)(x_i - a_i) = 0, \quad j = 1,\dots,r$$

is called the tangent space at $x = (a_1,\dots,a_n)$. (A more intrinsic definition is given by Zariski tangent space.) If the tangent space at x and the variety X have the same dimension, the point x is said to be smooth; otherwise, singular.

The important difference from calculus is that the inverse function theorem fails. To alleviate this problem, one has to consider the étale topology instead of the Zariski topology. (cf. Milne, Étale)

Projective Variety

In algebraic geometry, a projective variety over an algebraically closed field k is a subset of some projective n-space P^n over k that is the zero-locus of some finite family of homogeneous polynomials of $n + 1$ variables with coefficients in k, that generate a prime ideal, the defining ideal of the variety. Equivalently, an algebraic variety is projective if it can be embedded as a Zariski closed subvariety of P^n.

A projective variety is a projective curve if its dimension is one; it is a projective surface if its dimension is two; it is a projective hypersurface if its dimension is one less than the dimension of the containing projective space; in this case it is the set of zeros of a single homogeneous polynomial.

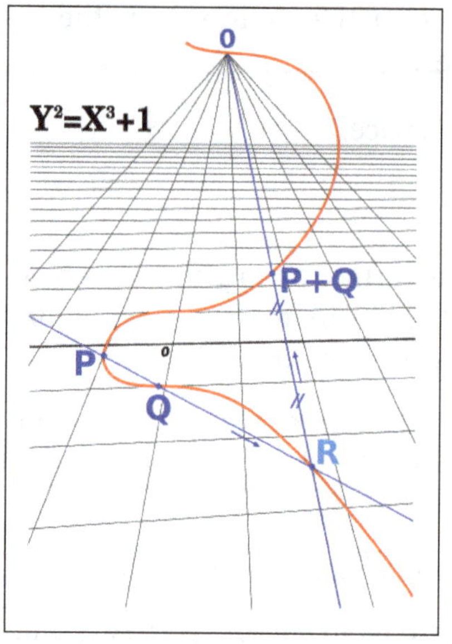

An elliptic curve is a smooth projective curve of genus one.

If X is a projective variety defined by a homogeneous prime ideal I, then the quotient ring

$$k[x_0,\ldots,x_n]/I$$

is called the homogeneous coordinate ring of X. Basic invariants of X such as the degree and the dimension can be read off the Hilbert polynomial of this graded ring.

Projective varieties arise in many ways. They are complete, which roughly can be expressed by saying that there are no points "missing". The converse is not true in general, but Chow's lemma describes the close relation of these two notions. Showing that a variety is projective is done by studying line bundles or divisors on X.

A salient feature of projective varieties are the finiteness constraints on sheaf cohomology. For smooth projective varieties, Serre duality can be viewed as an analog of Poincaré duality. It also leads to the Riemann-Roch theorem for projective curves, i.e., projective varieties of dimension 1. The theory of projective curves is particularly rich, including a classification by the genus of the curve. The classification program for higher-dimensional projective varieties naturally leads to the construction of moduli of projective varieties. Hilbert schemes parametrize closed subschemes of P^n with prescribed Hilbert polynomial. Hilbert schemes, of which Grassmannians are special cases, are also projective schemes in their own right. Geometric invariant theory offers another approach. The classical approaches include the Teichmüller space and Chow varieties.

A particularly rich theory, reaching back to the classics, is available for complex projective varieties, i.e., when the polynomials defining X have complex coefficients. Broadly, the GAGA principle says that the geometry of projective complex analytic spaces (or manifolds) is equivalent to the geometry of projective complex varieties. For example, the theory of holomorphic vector bundles (more generally coherent analytic sheaves) on X coincide with that of algebraic vector bundles. Chow's theorem says that a subset of projective space is the zero-locus of a family of holomorphic functions if and only if it is the zero-locus of homogeneous polynomials. The combination of analytic and algebraic methods for complex projective varieties lead to areas such as Hodge theory.

Variety and Scheme Structure

Variety Structure

Let k be an algebraically closed field. The basis of the definition of projective varieties is projective space P^n, which can be defined in different, but equivalent ways:

- as the set of all lines through the origin in k^{n+1} (i.e., one-dimensional sub-vector spaces of k^{n+1}).
- as the set of tuples $(x_0,\ldots,x_n) \in k^{n+1}$, modulo the equivalence relation

  $$(x_0,\ldots,x_n) \sim \lambda(x_0,\ldots,x_n).$$

  for any $\lambda \in k \setminus \{0\}$. The equivalence class of such a tuple is denoted by

  $$[x_0 : \cdots : x_n]$$

  and referred to as a homogeneous coordinate.

A *projective variety* is, by definition, a closed subvariety of P^n, where closed refers to the Zariski topology. In general, closed subsets of the Zariski topology are defined to be the zero-locus of polynomial functions. Given a polynomial $f \in k[x_0,\ldots,x_n]$, the condition;

$$f([x_0 : \cdots : x_n]) = 0$$

does not make sense for arbitrary polynomials, but only if f is homogeneous, i.e., the total degree of all the monomials (whose sum is f) is the same. In this case, the vanishing of:

$$f(\lambda x_0,\ldots,\lambda x_n) = \lambda^{\deg f} f(x_0,\ldots,x_n)$$

is independent of the choice of $\lambda(\neq 0)$.

Therefore, projective varieties arise from homogeneous prime ideals I of $k[x_0,\ldots,x_n]$, and setting;

$$X = \{[x_0 : \cdots : x_n] \in \mathbf{P}^n, f([x_0 : \cdots : x_n]) = 0 \text{ for all } f \in I\}.$$

Moreover, the projective variety X is an algebraic variety, meaning that it is covered by open affine subvarieties and satisfies the separation axiom. Thus, the local study of X (e.g., singularity) reduces to that of an affine variety. The explicit structure is as follows. The projective space P^n is covered by the standard open affine charts:

$$U_i = \{[x_0 : \cdots : x_n], x_i \neq 0\},$$

which themselves are affine n-spaces with the coordinate ring $k[y_1^{(i)},\ldots,y_n^{(i)}], y_j^{(i)} = x_j / x_i$. Say $i = 0$ for the notational simplicity and drop the superscript (0). Then $X \cap U_0$ is a closed subvariety of $U_0 \simeq \mathbb{A}^n$ defined by the ideal of $k[y_1,\ldots,y_n]$ generated by:

$$f(1, y_1,\ldots,y_n)$$

for all f in I. Thus, X is an algebraic variety covered by (n+1) open affine charts $X \cap U_i$.

Note that X is the closure of the affine variety $X \cap U_0$ in $\mathbb{P}^n$. Conversely, starting from some closed (affine) variety $V \subset U_0 \simeq \mathbb{A}^n$, the closure of V in $\mathbb{P}^n$ is the projective variety called the projective completion of V. If $I \subset k[y_1,\ldots,y_n]$ defines V, then the defining ideal of this closure is the homogeneous ideal of $k[x_0,\ldots,x_n]$ generated by:

$$x_0^{\deg(f)} f(x_1/x_0,\ldots,x_n/x_0)$$

for all f in I.

For example, if V is an affine curve given by, say, $y^2 = x^3 + ax + b$ in the affine plane, then its projective completion in the projective plane is given by $y^2 z = x^3 + axz^2 + bz^3$.

Projective Schemes

For various applications, it is necessary to consider more general algebro-geometric objects than projective varieties, namely projective schemes. The first step towards projective schemes is to endow projective space with a scheme structure, in a way refining the above description of projective space as an algebraic variety, i.e., $P^n(k)$ is a scheme which it is a union of (n + 1) copies of the affine n-space k^n. More generally, projective space over a ring A is the union of the affine schemes;

$$U_i = \operatorname{Spec} A[x_1/x_i,\ldots,x_n/x_i], \quad 0 \le i \le n,$$

in such a way the variables match up as expected. The set of closed points of $\mathbf{P}_k^n$, for algebraically closed fields k, is then the projective space $P^n(k)$ in the usual sense.

An equivalent but streamlined construction is given by the Proj construction, which is an analog of the spectrum of a ring, denoted “Spec”, which defines an affine scheme. For example, if A is a ring, then:

$$\mathbf{P}_A^n = \operatorname{Proj} A[x_0,\ldots,x_n].$$

If R is a quotient of $k[x_0,\ldots,x_n]$ by a homogeneous ideal I, then the canonical surjection induces the closed immersion:

$$\operatorname{Proj} R \hookrightarrow \mathbf{P}_k^n.$$

Compared to projective varieties, the condition that the ideal I be a prime ideal was dropped. This leads to a much more flexible notion: on the one hand the topological space $X = \operatorname{Proj} R$ may have multiple irreducible components. Moreover, there may be nilpotent functions on X.

Closed subschemes of $\mathbf{P}_k^n$ correspond bijectively to the homogeneous ideals I of $k[x_0,\ldots,x_n]$ that are saturated; i.e., $I:(x_0,\ldots,x_n) = I$. This fact may be considered as a refined version of projective Nullstellensatz.

We can give a coordinate-free analog of the above. Namely, given a finite-dimensional vector space V over k, we let:

$$\mathbf{P}(V) = \operatorname{Proj} k[V]$$

where $k[V] = \operatorname{Sym}(V^*)$ is the symmetric algebra of V^*. It is the projectivization of *V*; i.e., it parametrizes lines in *V*. There is a canonical surjective map $\pi : V - 0 \to \mathbf{P}(V)$, which is defined using the chart described above. One important use of the construction is this (cf., Duality and linear system). A divisor *D* on a projective variety *X* corresponds to a line bundle *L*. One then set:

$$|D| = \mathbf{P}(\Gamma(X, L));$$

it is called the complete linear system of *D*.

Projective space over any scheme *S* can be defined as a fiber product of schemes:

$$\mathbf{P}_S^n = \mathbf{P}_{\mathbf{Z}}^n \times_{\operatorname{Spec} \mathbf{Z}} S.$$

If $\mathcal{O}(1)$ is the twisting sheaf of Serre on $\mathbf{P}_{\mathbf{Z}}^n$, we let $\mathcal{O}(1)$ denote the pullback of $\mathcal{O}(1)$ to $\mathbf{P}_S^n$; that is, $\mathcal{O}(1) = g^*(\mathcal{O}(1))$ for the canonical map $g : \mathbf{P}_S^n \to \mathbf{P}_{\mathbf{Z}}^n$.

A scheme $X \to S$ is called projective over *S* if it factors as a closed immersion:

$$X \to \mathbf{P}_S^n$$

followed by the projection to *S*.

A line bundle (or invertible sheaf) $\mathcal{L}$ on a scheme *X* over *S* is said to be very ample relative to *S* if there is an immersion (i.e., an open immersion followed by a closed immersion);

$$i : X \to \mathbf{P}_S^n$$

for some *n* so that $\mathcal{O}(1)$ pullbacks to $\mathcal{L}$. Then a *S*-scheme *X* is projective if and only if it is proper and there exists a very ample sheaf on *X* relative to *S*. Indeed, if *X* is proper, then an immersion corresponding to the very ample line bundle is necessarily closed. Conversely, if *X* is projective, then the pullback of $\mathcal{O}(1)$ under the closed immersion of *X* into a projective space is very ample. That "projective" implies "proper" is deeper: the *main theorem of elimination theory*.

Relation to Complete Varieties

By definition, a variety is complete, if it is proper over *k*. The valuative criterion of properness expresses the intuition that in a proper variety, there are no points "missing".

There is a close relation between complete and projective varieties: on the one hand, projective space and therefore any projective variety is complete. The converse is not true in general. However:

- A smooth curve *C* is projective if and only if it is complete. This is proved by identifying *C* with the set of discrete valuation rings of the function field *k*(*C*) over *k*. This set has a natural Zariski topology called the Zariski–Riemann space.

- Chow's lemma states that for any complete variety *X*, there is a projective variety *Z* and a birational morphism $Z \to X$. (Moreover, through normalization, one can assume this projective variety is normal.)

Some properties of a projective variety follow from completeness. For example,

$$\Gamma(X, \mathcal{O}_X) = k$$

for any projective variety *X* over *k*. This fact is an algebraic analogue of Liouville's theorem (any holomorphic function on a connected compact complex manifold is constant). In fact, the similarity between complex analytic geometry and algebraic geometry on complex projective varieties goes much further than this, as is explained below.

Quasi-projective varieties are, by definition, those which are open subvarieties of projective varieties. This class of varieties includes affine varieties. Affine varieties are almost never complete (or projective). In fact, a projective subvariety of an affine variety must have dimension zero. This is because only the constants are globally regular functions on a projective variety.

Examples and Basic Invariants

By definition, any homogeneous ideal in a polynomial ring yields a projective scheme (required to be prime ideal to give a variety). In this sense, examples of projective varieties abound. The following list mentions various classes of projective varieties which are noteworthy since they have been studied particularly intensely. The important class of complex projective varieties, i.e., the case $k = \mathbf{C}$, is discussed further below.

The product of two projective spaces is projective. In fact, there is the explicit immersion (called Segre embedding);

$$\mathbf{P}^n \times \mathbf{P}^m \to \mathbf{P}^{(n+1)(m+1)-1}, (x_i, y_j) \mapsto x_i y_j.$$

As a consequence, the product of projective varieties over *k* is again projective. The Plücker embedding exhibits a Grassmannian as a projective variety. Flag varieties such as the quotient of the general linear group $GL_n(k)$ modulo the subgroup of upper triangular matrices, are also projective, which is an important fact in the theory of algebraic groups.

Homogeneous Coordinate Ring and Hilbert Polynomial

As the prime ideal *P* defining a projective variety *X* is homogeneous, the homogeneous coordinate ring;

$$R = k[x_0, \ldots, x_n] / P$$

is a graded ring, i.e., can be expressed as the direct sum of its graded components:

$$R = \bigoplus_{n \in \mathbf{N}} R_n.$$

There exists a polynomial P such that $\dim R_n = P(n)$ for all sufficiently large n; it is called the Hilbert polynomial of X. It is a numerical invariant encoding some extrinsic geometry of X. The degree of P is the dimension r of X and its leading coefficient times r! is the degree of the variety X. The arithmetic genus of X is $(-1)^r (P(0) - 1)$ when X is smooth.

For example, the homogeneous coordinate ring of $\mathbf{P}^n$ is $k[x_0,\dots,x_n]$ and its Hilbert polynomial is $P(z) = \binom{z+n}{n}$; its arithmetic genus is zero.

If the homogeneous coordinate ring R is an integrally closed domain, then the projective variety X is said to be projectively normal. Note, unlike normality, projective-normality depends on R, the embedding of X into a projective space. The normalization of a projective variety is projective; in fact, it's the Proj of the integral closure of some homogeneous coordinate ring of X.

Degree

Let $X \subset \mathbb{P}^N$ be a projective variety. There are at least two equivalent ways to define the degree of X relative to its embedding. The first way is to define it as the cardinality of the finite set;

$$\#(X \cap H_1 \cap \cdots \cap H_d)$$

where d is the dimension of X and H_i's are hyperplanes in "general positions". This definition corresponds to an intuitive idea of a degree. Indeed, if X is a hypersurface, then the degree of X is the degree of the homogeneous polynomial defining X. The "general positions" can be made precise, for example, by intersection theory; one requires that the intersection is proper and that the multiplicities of irreducible components are all one.

The other definition, which is mentioned in the previous section, is that the degree of X is the leading coefficient of the Hilbert polynomial of X times (dim X)!. Geometrically, this definition means that the degree of X is the multiplicity of the vertex of the affine cone over X.

Let $V_1,\dots,V_r \subset \mathbb{P}^N$ be closed subschemes of pure dimensions that intersect properly (they are in general position). If m_i denotes the multiplicity of an irreducible component Z_i in the intersection (i.e., intersection multiplicity), then the generalization of Bézout's theorem says:

$$\sum_1^s m_i \deg Z_i = \prod_1^r \deg V_i .$$

The intersection multiplicity m_i can be defined as the coefficient of Z_i in the intersection product $V_1 \cdot \ldots \cdot V_r$ in the Chow ring of $\mathbb{P}^N$.

In particular, if $H \subset \mathbb{P}^N$ is a hypersurface not containing X, then;

$$\sum_1^s m_i \deg Z_i = \deg(X)\deg(H)$$

where Z_i's are the irreducible components of the scheme-theoretic intersection of X and H with multiplicity (length of the local ring) m_i.

The Ring of Sections

Let X be a projective variety and L a line bundle on it. Then the graded ring;

$$R(X,L) = \bigoplus_{n=0}^{\infty} H^0(X, L^{\otimes n})$$

is called the ring of sections of L. If L is ample, then Proj of this ring is X. Moreover, if X is normal and L is very ample, then $R(X,L)$ is the integral closure of the homogeneous coordinate ring of X determined by L; i.e., $X \hookrightarrow \mathbb{P}^N$ so that $\mathcal{O}_{\mathbb{P}^N}(1)$ pulls-back to L.

For applications, it is useful to allow for divisors (or $\mathbb{Q}$-divisors) not just line bundles; assuming X is normal, the resulting ring is then called a generalized ring of sections. If K_X is a canonical divisor on X, then the generalized ring of sections;

$$R(X, K_X)$$

is called the canonical ring of X. If the canonical ring is finitely generated, then Proj of the ring is called the canonical model of X. The canonical ring or model can then be used to define the Kodaira dimension of X.

Projective Curves

Projective schemes of dimension one are called *projective curves*. Much of the theory of projective curves is about smooth projective curves, since the singularities of curves can be resolved by normalization, which consists in taking locally the integral closure of the ring of regular functions. Smooth projective curves are isomorphic if and only if their function fields are isomorphic. The study of finite extensions of:

$$\mathbb{F}_p(t),$$

or equivalently smooth projective curves over $\mathbb{F}_p$ is an important branch in algebraic number theory.

A smooth projective curve of genus one is called an elliptic curve. As a consequence of the Riemann-Roch theorem, such a curve can be embedded as a closed subvariety in $\mathbf{P}^2$. In general, any (smooth) projective curve can be embedded in $\mathbf{P}^3$. Conversely, any smooth closed curve in $\mathbf{P}^2$ of degree three has genus one by the genus formula and is thus an elliptic curve.

A smooth complete curve of genus greater than or equal to two is called a hyperelliptic curve if there is a finite morphism $C \to \mathbf{P}^1$ of degree two.

Projective Hypersurfaces

Every irreducible closed subset of $\mathbf{P}^n$ of codimension one is a hypersurface; i.e., the zero set of some homogeneous irreducible polynomial.

Abelian Varieties

Another important invariant of a projective variety X is the Picard group $\mathrm{Pic}(X)$ of X, the set of isomorphism classes of line bundles on X. It is isomorphic to $H^1(X,\mathcal{O}_X^*)$ and therefore an intrinsic notion (independent of embedding). For example, the Picard group of $\mathbb{P}^n$ is isomorphic to Z via the degree map. The kernel of $\deg:\mathrm{Pic}(X)\to\mathbf{Z}$ is not only an abstract abelian group, but there is a variety called the Jacobian variety of X, Jac(X), whose points equal this group. The Jacobian of a (smooth) curve plays an important role in the study of the curve. For example, the Jacobian of an elliptic curve E is E itself. For a curve X of genus g, Jac(X) has dimension g.

Varieties, such as the Jacobian variety, which are complete and have a group structure are known as abelian varieties, in honor of Niels Abel. In marked contrast to affine algebraic groups such as $GL_n(k)$, such groups are always commutative, whence the name. Moreover, they admit an ample line bundle and are thus projective. On the other hand, an abelian scheme may not be projective. Examples of abelian varieties are elliptic curves, Jacobian varieties and K3 surfaces.

Projections

Let $E\subset\mathbb{P}^n$ be a linear subspace; i.e., $E=\{s_0=s_1=\ldots=s_r=0\}$ for some linearly independent linear functional s_i's. Then the projection from $\boldsymbol{E}$ is the (well-defined) morphism;

$$\phi:\mathbb{P}^n-E\to\mathbb{P}^r, x\mapsto[s_0(x):\cdots:s_r(x)].$$

- The geometric description of this map is the following. We view $\mathbb{P}^r\subset\mathbb{P}^n$ so that it is disjoint from E. Then, for any $x\in\mathbb{P}^n-E$,

  $$\phi(x)=W_x\cap\mathbb{P}^r$$

  where we wrote W_x for the smallest linear space containing E and x (called the join of E and x.)
- $\phi^{-1}(\{y_i\neq 0\})=\{s_i\neq 0\}$, where y_i are the homogeneous coordinates on $\mathbb{P}^r$.
- For any closed subscheme $Z\subset\mathbb{P}^n$ disjoint from E, the restriction;

  $$\phi:Z\to\mathbb{P}^r$$

  is a finite morphism.

Projections can be used to cut down the dimension in which a projective variety is embedded, up to finite morphisms. Start with some projective variety $X\subset\mathbb{P}^n$. If $n>\dim X$, the projection from a point not on X gives $\phi:X\to\mathbb{P}^{n-1}$. Moreover, ϕ is a finite map to its image. Thus, iterating the procedure, one sees there is a finite map;

$$X\to\mathbb{P}^d, d=\dim X.$$

This result is the projective analog of Noether's normalization lemma. (In fact, it yields a geometric proof of the normalization lemma.)

The same procedure can be used to show the following slightly more precise result: given a projective variety X over a perfect field, there is a finite birational morphism from X to a hypersurface H in $\mathbb{P}^{d+1}$. In particular, if X is normal, then it is the normalization of H.

Duality and Linear System

While a projective n-space $\mathbb{P}^n$ parameterizes the lines in an affine n-space, the dual of it parametrizes the hyperplanes on the projective space, as follows. Fix a field k. By $\check{\mathbb{P}}^n_k$, we mean a projective n-space;

$$\check{\mathbb{P}}^n_k = \operatorname{Proj}(k[u_0,\dots,u_n])$$

equipped with the construction:

$$f \mapsto H_f = \{\alpha_0 x_0 + \cdots + \alpha_n x_n = 0\},\ \text{a hyperplane on } \mathbb{P}^n_L$$

where $f : \operatorname{Spec} L \to \check{\mathbb{P}}^n_k$ is an L-point of $\check{\mathbb{P}}^n_k$ for a field extension L of k and $\alpha_i = f^*(u_i) \in L$.

For each L, the construction is a bijection between the set of L-points of $\check{\mathbb{P}}^n_k$ and the set of hyperplanes on $\mathbb{P}^n_L$. Because of this, the dual projective space $\check{\mathbb{P}}^n_k$ is said to be the moduli space of hyperplanes on .

A line in $\check{\mathbb{P}}^n_k$ is called a pencil: it is a family of hyperplanes on $\mathbb{P}^n_k$ parametrized by $\mathbb{P}^1_k$.

If V is a finite-dimensional vector space over k, then, for the same reason as above, $\mathbb{P}(V^*) = \operatorname{Proj}(\operatorname{Sym}(V))$ is the space of hyperplanes on $\mathbb{P}(V)$. An important case is when V consists of sections of a line bundle. Namely, let X be an algebraic variety, L a line bundle on X and $V \subset \Gamma(X, L)$ a vector subspace of finite positive dimension. Then there is:

$$\varphi_V : X - B \to \mathbb{P}(V^*)$$

given by

$$x \mapsto H_x = \{s \in V \mid s(x) = 0\}$$

and called the map determined by the linear system V, where B, called the base locus, is the intersection of the divisors of zero of nonzero sections in V.

Cohomology of Coherent Sheaves

Let X be a projective scheme over a field (or, more generally over a Noetherian ring A). Cohomology of coherent sheaves $\mathcal{F}$ on X satisfies the following important theorems due to Serre:

1. $H^p(X, \mathcal{F})$ is a finite-dimensional k-vector space for any p.
2. There exists an integer n_0 (depending on $\mathcal{F}$; Castelnuovo–Mumford regularity) such that

   $H^p(X, \mathcal{F}(n)) = 0$.

for all $n \geq n_0$ and $p > 0$, where $\mathcal{F}(n) = \mathcal{F} \otimes \mathcal{O}(n)$ is the twisting with a power of a very ample line bundle $\mathcal{O}(1)$

These results are proven reducing to the case $X = \mathbf{P}^n$ using the isomorphism

$$H^p(X, \mathcal{F}) = H^p(\mathbf{P}^r, \mathcal{F}), p \geq 0$$

where in the right-hand side $\mathcal{F}$ is viewed as a sheaf on the projective space by extension by zero. The result then follows by a direct computation for $\mathcal{F} = \mathcal{O}_{\mathbf{P}^r}(n)$, *n* any integer, and for arbitrary $\mathcal{F}$ reduces to this case without much difficulty.

As a corollay to 1. above, if *f* is a projective morphism from a noetherian scheme to a noetherian ring, then the higher direct image $R^p f_* \mathcal{F}$ is coherent. The same result holds for proper morphisms *f*, as can be shown with the aid of Chow's lemma.

Sheaf cohomology groups H^i on a noetherian topological space vanish for *i* strictly greater than the dimension of the space. Thus the quantity, called the Euler characteristic of $\mathcal{F}$,

$$\chi(\mathcal{F}) = \sum_{i=0}^{\infty} (-1)^i \dim H^i(X, \mathcal{F})$$

is a well-defined integer (for *X* projective). One can then show $\chi(\mathcal{F}(n)) = P(n)$ for some polynomial *P* over rational numbers. Applying this procedure to the structure sheaf $\mathcal{O}_X$, one recovers the Hilbert polynomial of *X*. In particular, if *X* is irreducible and has dimension *r*, the arithmetic genus of *X* is given by:

$$(-1)^r (\chi(\mathcal{O}_X) - 1),$$

which is manifestly intrinsic; i.e., independent of the embedding.

The arithmetic genus of a hypersurface of degree *d* is $\binom{d-1}{n}$ in $\mathbf{P}^n$. In particular, a smooth curve of degree *d* in P^2 has arithmetic genus $(d-1)(d-2)/2$. This is the genus formula.

Smooth Projective Varieties

Let *X* be a smooth projective variety where all of its irreducible components have dimension *n*. In this situation, the canonical sheaf ω_X, defined as the sheaf of Kähler differentials of top degree (i.e., algebraic *n*-forms), is a line bundle.

Serre Duality

Serre duality states that for any locally free sheaf $\mathcal{F}$ on *X*,

$$H^i(X, \mathcal{F}) \simeq H^{n-i}(X, \mathcal{F}^\vee \otimes \omega_X)';$$

where the superscript prime refers to the dual space and $\mathcal{F}^\vee$ is the dual sheaf of $\mathcal{F}$. A generalization to projective, but not necessarily smooth schemes is known as Verdier duality.

Riemann-Roch Theorem

For a (smooth projective) curve X, H^2 and higher vanish for dimensional reason and the space of the global sections of the structure sheaf is one-dimensional. Thus the arithmetic genus of X is the dimension of $H^1(X,\mathcal{O}_X)$. By definition, the geometric genus of X is the dimension of $H^0(X, \omega_X)$. Serre duality thus implies that the arithmetic genus and the geometric genus coincide. They will simply be called the genus of X.

Serre duality is also a key ingredient in the proof of the Riemann–Roch theorem. Since X is smooth, there is an isomorphism of groups;

$$\mathrm{Cl}(X) \to \mathrm{Pic}(X), D \mapsto \mathcal{O}(D)$$

from the group of (Weil) divisors modulo principal divisors to the group of isomorphism classes of line bundles. A divisor corresponding to ω_X is called the canonical divisor and is denoted by K. Let $l(D)$ be the dimension of $H^0(X,\mathcal{O}(D))$. Then the Riemann–Roch theorem states: if g is a genus of X,

$$l(D) - l(K - D) = \deg D + 1 - g$$

for any divisor D on X. By the Serre duality, this is the same as:

$$\chi(\mathcal{O}(D)) = \deg D + 1 - g\,,$$

which can be readily proved. A generalization of the Riemann-Roch theorem to higher dimension is the Hirzebruch-Riemann-Roch theorem, as well as the far-reaching Grothendieck-Riemann-Roch theorem.

Hilbert Schemes

Hilbert schemes parametrize all closed subvarieties of a projective scheme X in the sense that the points (in the functorial sense) of H correspond to the closed subschemes of X. As such, the Hilbert scheme is an example of a moduli space, i.e., a geometric object whose points parametrize other geometric objects. More precisely, the Hilbert scheme parametrizes closed subvarieties whose Hilbert polynomial equals a prescribed polynomial P. It is a deep theorem of Grothendieck that there is a scheme H_X^P over k such that, for any k-scheme T, there is a bijection;

$$\{\mathit{morphisms}\ T \to H_X^P\} \longleftrightarrow \{\mathit{closed\ subschemes\ of}\ X \times_k T\ \mathit{flat\ over}\ T,\ \mathit{with\ Hilbert\ polynomial}\ P.\}$$

The closed subscheme of $X \times H_X^P$ that corresponds to the identity map $H_X^P \to H_X^P$ is called the *universal family*.

For $P(z) = \binom{z+r}{r}$, the Hilbert scheme $H_{\mathbf{P}^n}^P$ is called the Grassmannian of r-planes in $\mathbf{P}^n$ and, if X is a projective scheme, H_X^P is called the Fano scheme of r-planes on X.

Complex Projective Varieties

In this section, all algebraic varieties are complex algebraic varieties. A key feature of the theory of complex projective varieties is the combination of algebraic and analytic methods. The transition between these theories is provided by the following link: since any complex polynomial is also a holomorphic function, any complex variety *X* yields a complex analytic space, denoted $X(\mathbf{C})$. Moreover, geometric properties of *X* are reflected by the ones of $X(\mathbf{C})$. For example, the latter is a complex manifold iff *X* is smooth; it is compact iff *X* is proper over C.

Relation to Complex Kähler Manifolds

Complex projective space is a Kähler manifold. This implies that, for any projective algebraic variety *X*, *X*(C) is a compact Kähler manifold. The converse is not in general true, but the Kodaira embedding theorem gives a criterion for a Kähler manifold to be projective.

In low dimensions, there are the following results:

- (Riemann) A compact Riemann surface (i.e., compact complex manifold of dimension one) is a projective variety. By the Torelli theorem, it is uniquely determined by its Jacobian.
- (Chow-Kodaira) A compact complex manifold of dimension two with two algebraically independent meromorphic functions is a projective variety.

GAGA and Chow's Theorem

Chow's theorem provides a striking way to go the other way, from analytic to algebraic geometry. It states that every analytic subvariety of a complex projective space is algebraic. The theorem may be interpreted to saying that a holomorphic function satisfying certain growth condition is necessarily algebraic: "projective" provides this growth condition. One can deduce from the theorem the following:

- Meromorphic functions on the complex projective space are rational.
- If an algebraic map between algebraic varieties is an analytic isomorphism, then it is an (algebraic) isomorphism. (This part is a basic fact in complex analysis.) In particular, Chow's theorem implies that a holomorphic map between projective varieties is algebraic. (consider the graph of such a map.)
- Every holomorphic vector bundle on a projective variety is induced by a unique algebraic vector bundle.
- Every holomorphic line bundle on a projective variety is a line bundle of a divisor.

Chow's theorem can be shown via Serre's GAGA principle. Its main theorem states:

> Let *X* be a projective scheme over C. Then the functor associating the coherent sheaves on *X* to the coherent sheaves on the corresponding complex analytic space X^{an} is an equivalence of categories. Furthermore, the natural maps;
>
> $$H^i(X,\mathcal{F}) \to H^i(X^{an},\mathcal{F})$$

are isomorphisms for all *i* and all coherent sheaves $\mathcal{F}$ on *X*.

Complex Tori vs. Complex Abelian Varieties

The complex manifold associated to an abelian variety A over C is a compact complex Lie group. These can be shown to be of the form:

$$\mathbb{C}^g / L$$

and are also referred to as complex tori. Here, g is the dimension of the torus and L is a lattice (also referred to as period lattice).

According to the uniformization theorem already mentioned above, any torus of dimension 1 arises from an abelian variety of dimension 1, i.e., from an elliptic curve. In fact, the Weierstrass's elliptic function $\wp$ attached to L satisfies a certain differential equation and as a consequence it defines a closed immersion:

$$\mathbb{C}/L \to \mathbf{P}^2, L \mapsto (0:0:1), z \mapsto (1:\wp(z):\wp'(z)).$$

There is a p-adic analog, the p-adic uniformization theorem.

For higher dimensions, the notions of complex abelian varieties and complex tori differ: only polarized complex tori come from abelian varieties.

Kodaira Vanishing

The fundamental Kodaira vanishing theorem states that for an ample line bundle $\mathcal{L}$ on a smooth projective variety X over a field of characteristic zero,

$$H^i(X, \mathcal{L} \otimes \omega_X) = 0$$

for $i > 0$, or, equivalently by Serre duality $H^i(X, \mathcal{L}^{-1}) = 0$ for $i < n$. The first proof of this theorem used analytic methods of Kähler geometry, but a purely algebraic proof was found later. The Kodaira vanishing in general fails for a smooth projective variety in positive characteristic. Kodaira's theorem is one of various vanishing theorems, which give criteria for higher sheaf cohomologies to vanish. Since the Euler characteristic of a sheaf is often more manageable than individual cohomology groups, this often has important consequences about the geometry of projective varieties.

Related Notions

- Multi-projective variety.
- *Weighted projective variety*, a closed subvariety of a weighted projective space.

Quasi-projective Variety

In mathematics, a quasi-projective variety in algebraic geometry is a locally closed subset of a projective variety, i.e., the intersection inside some projective space of a Zariski-open and a Zariski-closed subset. A similar definition is used in scheme theory, where a *quasi-projective scheme* is a locally closed subscheme of some projective space.

Relationship to Affine Varieties

An affine space is a Zariski-open subset of a projective space, and since any closed affine subset U can be expressed as an intersection of the projective completion $\bar{U}$ and the affine space embedded in the projective space, this implies that any affine variety is quasiprojective. There are locally closed subsets of projective space that are not affine, so that quasi-projective is more general than affine. Taking the complement of a single point in projective space of dimension at least 2 gives a non-affine quasi-projective variety. This is also an example of a quasi-projective variety that is neither affine nor projective.

Examples

Since quasi-projective varieties generalize both affine and projective varieties, they are sometimes referred to simply as *varieties*. Varieties isomorphic to affine algebraic varieties as quasi-projective varieties are called affine varieties; similarly for projective varieties. For example, the complement of a point in the affine line, i.e., $X = \mathbb{A}^1 \setminus \{0\}$, is isomorphic to the zero set of the polynomial $xy-1$ in the affine plane. As an affine set X is not closed since any polynomial zero on the complement must be zero on the affine line. For another example, the complement of any conic in projective space of dimension 2 is affine. Varieties isomorphic to open subsets of affine varieties are called quasi-affine.

Quasi-projective varieties are *locally affine* in the sense that a manifold is locally Euclidean — every point of a quasiprojective variety has a neighborhood given by an affine variety. This yields a basis of affine sets for the Zariski topology on a quasi-projective variety.

Hilbert's Nullstellensatz

Hilbert's Nullstellensatz (German for "theorem of zeros," or more literally, "zero-locus-theorem") is a theorem that establishes a fundamental relationship between geometry and algebra. This relationship is the basis of algebraic geometry, a branch of mathematics. It relates algebraic sets to ideals in polynomial rings over algebraically closed fields. This relationship was discovered by David Hilbert who proved the Nullstellensatz and several other important related theorems named after him (like Hilbert's basis theorem).

Formulation

Let k be a field (such as the rational numbers) and K be an algebraically closed field extension (such as the complex numbers), consider the polynomial ring $k[X_1,X_2,\ldots, X_n]$ and let I be an ideal in this ring. The algebraic set $V(I)$ defined by this ideal consists of all n-tuples $x = (x_1,\ldots,x_n)$ in K^n such that $f(x) = 0$ for all f in I. Hilbert's Nullstellensatz states that if p is some polynomial in $k[X_1,X_2,\ldots, X_n]$ that vanishes on the algebraic set $V(I)$, i.e. $p(x) = 0$ for all x in $V(I)$, then there exists a natural number r such that p^r is in I.

An immediate corollary is the "weak Nullstellensatz": The ideal I in $k[X_1,X_2,\ldots, X_n]$ contains 1 if and only if the polynomials in I do not have any common zeros in K^n. It may also be formulated as follows: if I is a proper ideal in $k[X_1,X_2,..., X_n]$, then V(I) cannot be empty, i.e. there exists a common zero for all the polynomials in the ideal in every algebraically closed extension of k. This is the reason for the name of the theorem, which can be proved easily from the 'weak' form using the Rabinowitsch trick. The assumption of considering common zeros in an algebraically closed field is essential here; for example, the elements of the proper ideal $(X^2 + 1)$ in R[X] do not have a common zero in R. With the notation common in algebraic geometry, the Nullstellensatz can also be formulated as:

$$I(V(J)) = \sqrt{J}$$

for every ideal J. Here, $\sqrt{J}$ denotes the radical of J and I(U) is the ideal of all polynomials that vanish on the set U.

In this way, we obtain an order-reversing bijective correspondence between the algebraic sets in K^n and the radical ideals of $K[X_1,X_2,..., X_n]$. In fact, more generally, one has a Galois connection between subsets of the space and subsets of the algebra, where "Zariski closure" and "radical of the ideal generated" are the closure operators.

As a particular example, consider a point $P = (a_1,\ldots,a_n) \in K^n$. Then $I(P) = (X_1 - a_1,\ldots,X_n - a_n)$. More generally,

$$\sqrt{I} = \bigcap_{(a_1,\ldots,a_n)\in V(I)} (X_1 - a_1,\ldots,X_n - a_n).$$

Conversely, every maximal ideal of the polynomial ring $K[X_1,\ldots,X_n]$ (note that K is algebraically closed) is of the form $(X_1 - a_1,\ldots,X_n - a_n)$ for some $a_1,\ldots,a_n \in K$.

As another example, an algebraic subset W in K^n is irreducible (in the Zariski topology) if and only if $I(W)$ is a prime ideal.

Proof and Generalization

There are many known proofs of the theorem. One proof uses Zariski's lemma, which asserts that an associative algebra that is finitely generated over a field k is a field is a finite field extension of k.

Let $A = k[t_1,\ldots,t_n]$ (k algebraically closed field), I an ideal of A and V the common zeros of I in k^n. Clearly, $\sqrt{I} \subseteq I(V)$. Let $f \notin \sqrt{I}$. Then $f \notin \mathfrak{p}$ for some prime ideal $\mathfrak{p} \supseteq I$ in A. Let $R = (A/\mathfrak{p})[f^{-1}]$ and $\mathfrak{m}$ a maximal ideal in R. By Zariski's lemma, $R/\mathfrak{m}$ is a finite extension of k; thus, is k since k is algebraically closed. Let x_i be the images of t_i under the natural map $A \to k$. It follows that $x = (x_1,\ldots,x_n) \in V$ and $f(x) \neq 0$.

The Nullstellensatz will also follow trivially once one systematically developed the theory of a Jacobson ring, a ring in which a radical ideal is an intersection of maximal ideals. Let R be a Jacobson ring. If S is a finitely generated R-algebra, then S is a Jacobson ring. Further, if $\mathfrak{n} \subset S$ is a maximal ideal, then $\mathfrak{m} := \mathfrak{n} \cap R$ is a maximal ideal of R, and $S/\mathfrak{n}$ is a finite extension field of $R/\mathfrak{m}$.

Another generalization states that a faithfully flat morphism of schemes $f: Y \to X$ locally of finite type with X quasi-compact has a *quasi-section*, i.e. there exists X' affine and faithfully flat and quasi-finite over X together with an X-morphism $X' \to Y$.

Effective Nullstellensatz

In all of its variants, Hilbert's Nullstellensatz asserts that some polynomial g belongs or not to an ideal generated, say, by $f_1, \ldots, f_k$; we have $g = f^r$ in the strong version, $g = 1$ in the weak form. This means the existence or the non-existence of polynomials $g_1, \ldots, g_k$ such that $g = f_1 g_1 + \ldots + f_k g_k$. The usual proofs of the Nullstellensatz are not constructive, non-effective, in the sense that they do not give any way to compute the g_i.

It is thus a rather natural question to ask if there is an effective way to compute the g_i (and the exponent r in the strong form) or to prove that they do not exist. To solve this problem, it suffices to provide an upper bound on the total degree of the g_i: such a bound reduces the problem to a finite system of linear equations that may be solved by usual linear algebra techniques. Any such upper bound is called an effective Nullstellensatz.

A related problem is the ideal membership problem, which consists in testing if a polynomial belongs to an ideal. For this problem also, a solution is provided by an upper bound on the degree of the g_i. A general solution of the ideal membership problem provides an effective Nullstellensatz, at least for the weak form.

In 1925, Grete Hermann gave an upper bound for ideal membership problem that is doubly exponential in the number of variables. In 1982 Mayr and Meyer gave an example where the g_i have a degree that is at least double exponential, showing that every general upper bound for the ideal membership problem is doubly exponential in the number of variables.

Since most mathematicians at the time assumed the effective Nullstellensatz was at least as hard as ideal membership, few mathematicians sought a bound better than double-exponential. In 1987, however, W. Dale Brownawell gave an upper bound for the effective Nullstellensatz that is simply exponential in the number of variables. Brownawell's proof relied on analytic techniques valid only in characteristic 0, but, one year later, János Kollár gave a purely algebraic proof, valid in any characteristic, of a slightly better bound.

In the case of the weak Nullstellensatz, Kollár's bound is the following:

> Let $f_1, \ldots, f_s$ be polynomials in $n \geq 2$ variables, of total degree $d_1 \geq \ldots \geq d_s$. If there exist polynomials g_i such that $f_1 g_1 + \ldots + f_s g_s = 1$, then they can be chosen such that
>
> $$\deg(f_i g_i) \leq \max(d_s, 3) \prod_{j=1}^{\min(n,s)-1} \max(d_j, 3).$$
>
> This bound is optimal if all the degrees are greater than 2.

If d is the maximum of the degrees of the f_i, this bound may be simplified to

$$\max(3, d)^{\min(n,s)}.$$

Kollár's result has been improved by several authors. As of 14 October 2012, the best improvement, due to M. Sombra is:

$$\deg(f_i g_i) \le 2d_s \prod_{j=1}^{\min(n,s)-1} d_j.$$

His bound improves Kollár's as soon as at least two of the degrees that are involved are lower than 3.

Projective Nullstellensatz

We can formulate a certain correspondence between homogeneous ideals of polynomials and algebraic subsets of a projective space, called the projective Nullstellensatz, that is analogous to the affine one. To do that, we introduce some notations. Let $R = k[t_0,\ldots,t_n]$. The homogeneous ideal $R_+ = \bigoplus_{d\ge1} R_d$ is called the *maximal homogeneous ideal.* As in the affine case, we let: for a subset $S \subseteq \mathbb{P}^n$ and a homogeneous ideal *I* of *R*,

$$\mathrm{I}_{\mathbb{P}^n}(S) = \{f \in R_+ \mid f = 0 \text{ on } S\},$$
$$\mathrm{V}_{\mathbb{P}^n}(I) = \{x \in \mathbb{P}^n \mid f(x) = 0 \text{ for all } f \in I\}.$$

By $f = 0 \text{ on } S$ we mean: for every homogeneous coordinates $(a_0 : \cdots : a_n)$ of a point of *S* we have $f(a_0,\ldots,a_n) = 0$. This implies that the homogeneous components of *f* are also zero on *S* and thus that $\mathrm{I}_{\mathbb{P}^n}(S)$ is a homogeneous ideal. Equivalently, $\mathrm{I}_{\mathbb{P}^n}(S)$ is the homogeneous ideal generated by homogeneous polynomials *f* that vanish on *S*. Now, for any homogeneous ideal $I \subseteq R_+$, by the usual Nullstellensatz, we have:

$$\sqrt{I} = \mathrm{I}_{\mathbb{P}^n}(\mathrm{V}_{\mathbb{P}^n}(I)),$$

and so, like in the affine case, we have:

> There exists an order-reversing one-to-one correspondence between proper homogeneous radical ideals of *R* and subsets of $\mathbb{P}^n$ of the form $\mathrm{V}_{\mathbb{P}^n}(I)$. The correspondence is given by $\mathrm{I}_{\mathbb{P}^n}$ and $\mathrm{V}_{\mathbb{P}^n}$.

Analytic Nullstellensatz

The Nullstellensatz also holds for the germs of holomorphic functions at a point of complex *n*-space $\mathbb{C}^n$. Precisely, for each open subset $U \subset \mathbb{C}^n$, let $\mathcal{O}_{\mathbb{C}^n}(U)$ denote the ring of holomorphic functions on *U*; then $\mathcal{O}_{\mathbb{C}^n}$ is a *sheaf* on $\mathbb{C}^n$. The stalk $\mathcal{O}_{\mathbb{C}^n,0}$ at, say, the origin can be shown to be a Noetherian local ring that is a unique factorization domain.

If $f \in \mathcal{O}_{\mathbb{C}^n,0}$ is a germ represented by a holomorphic function $\tilde{f} : U \to \mathbb{C}$, then let

$V_0(f)$ = the equivalence class of the set $\{z \in U \mid \tilde{f}(z) = 0\}$

where two subsets *X*, *Y* of $\mathbb{C}^n$ are considered equivalent if $X \cap U = Y \cap U$ for some neighborhood *U* of 0. Note $V_0(f)$ is independent of a choice of the representative $\tilde{f}$. For each ideal $I \subset \mathcal{O}_{\mathbb{C}^n,0}$, let $V_0(I)$ denote $V_0(f_1) \cap \ldots \cap V_0(f_r)$ for some generators $f_1,\ldots,f_r$ of *I*. It is well-defined; i.e., is independent of a choice of the generators.

For each subset $X \subset \mathbb{C}^n$, let $I_0(X) = \{f \in \mathcal{O}_{\mathbb{C}^n,0} \mid V_0(f) \supset X\}$. It is easy to see that $I_0(X)$ is an ideal of $\mathcal{O}_{\mathbb{C}^n,0}$ and that $I_0(X) = I_0(Y)$ if $X \sim Y$ in the sense discussed above.

The analytic Nullstellensatz then states: for each ideal $I \subset \mathcal{O}_{\mathbb{C}^n,0}$,

$$\sqrt{I} = I_0(V_0(I))$$

where the left-hand side is the radical of *I*.

Zariski's Main Theorem

In algebraic geometry, Zariski's main theorem, proved by Oscar Zariski (1943), is a statement about the structure of birational morphisms stating roughly that there is only one branch at any normal point of a variety. It is the special case of Zariski's connectedness theorem when the two varieties are birational.

Zariski's main theorem can be stated in several ways which at first sight seem to be quite different, but are in fact deeply related. Some of the variations that have been called Zariski's main theorem are as follows:

- The total transform of a normal fundamental point of a birational map has positive dimension. This is essentially Zariski's original form of his main theorem.
- A birational morphism with finite fibers to a normal variety is an isomorphism to an open subset.
- The total transform of a normal point under a proper birational morphism is connected.
- A closely related theorem of Grothendieck describes the structure of quasi-finite morphisms of schemes, which implies Zariski's original main theorem.
- Several results in commutative algebra that imply the geometric form of Zariski's main theorem.
- A normal local ring is unibranch, which is a variation of the statement that the transform of a normal point is connected.
- The local ring of a normal point of a variety is analytically normal. This is a strong form of the statement that it is unibranch.

The name "Zariski's main theorem" comes from the fact that Zariski labelled it as the "MAIN THEOREM" in Zariski (1943).

Zariski's Main Theorem for Birational Morphisms

Let *f* be a birational mapping of algebraic varieties *V* and *W*. Recall that *f* is defined by a closed subvariety $\Gamma \subset V \times W$ (a "graph" of *f*) such that the projection on the first factor p_1 induces an isomorphism between an open $U \subset V$ and $p_1^{-1}(U)$, and such that $p_2 \circ p_1^{-1}$ is an isomorphism on *U* too.

The complement of U in V is called a *fundamental variety* or *indeteminancy locus*, and an image of a subset of V under $p_2 \circ p_1$ is called a *total transform* of it.

The original statement of the theorem reads:

> MAIN THEOREM: If W is an irreducible fundamental variety on V of a birational correspondence T between V and V' and if T has no fundamental elements on V' then — under the assumption that V is locally normal at W — each irreducible component of the transform $T[W]$ is of higher dimension than W.

Here T is essentially a morphism from V' to V that is birational, W is a subvariety of the set where the inverse of T is not defined whose local ring is normal, and the transform $T[W]$ means the inverse image of W under the morphism from V' to V.

Here are some variants of this theorem stated using more recent terminology. Hartshorne calls the following connectedness statement "Zariski's Main theorem":

> If $f:X\rightarrow Y$ is a birational projective morphism between noetherian integral schemes, then the inverse image of every normal point of Y is connected.

The following consequence of it also goes under this name:

> If $f:X\rightarrow Y$ is a birational transformation of projective varieties with Y normal, then the total transform of a fundamental point of f is connected and of dimension at least 1.

Examples

- Suppose that V is a smooth variety of dimension greater than 1 and V' is given by blowing up a point W on V. Then V is normal at W, and the component of the transform of W is a projective space, which has dimension greater than W as predicted by Zariski's original form of his main theorem.
- In the previous example the transform of W was irreducible. It is easy to find examples where the total transform is reducible by blowing up other points on the transform. For example, if V' is given by blowing up a point W on V and then blowing up another point on this transform, the total transform of W has 2 irreducible components meeting at a point. As predicted by Hartshorne's form of the main theorem, the total transform is connected and of dimension at least 1.
- For an example where W is not normal and the conclusion of the main theorem fails, take V' to be a smooth variety, and take V to be given by identifying two distinct points on V', and take W to be the image of these two points. Then W is not normal, and the transform of W consists of two points, which is not connected and does not have positive dimension.

Zariski's Main Theorem for Quasifinite Morphisms

In EGA III, Grothendieck calls the following statement which does not involve connectedness a "Main theorem" of Zariski Grothendieck:

> If $f:X \to Y$ is a quasi-projective morphism of Noetherian schemes then the set of points that are isolated in their fiber is open in X. Moreover the induced scheme of this set is isomorphic to an open subset of a scheme that is finite over Y.

In EGA IV, Grothendieck observed that the last statement could be deduced from a more general theorem about the structure of quasi-finite morphisms, and the latter is often referred to as the "Zariski's main theorem in the form of Grothendieck". It is well known that open immersions and finite morphisms are quasi-finite. Grothendieck proved that under the hypothesis of separatedness all quasi-finite morphisms are compositions of such Grothendieck:

> if Y is a quasi-compact separated scheme and $f : X \to Y$ is a separated, quasi-finite, finitely presented morphism then there is a factorization into $X \to Z \to Y$, where the first map is an open immersion and the second one is finite.

The relation between this theorem about quasi-finite morphisms and Théorème 4.4.3 of EGA III quoted above is that if $f:X \to Y$ is a projective morphism of varieties, then the set of points that are isolated in their fiber is quasifinite over Y. Then structure theorem for quasi-finite morphisms applies and yields the desired result.

Zariski's Main Theorem for Commutative Rings

Zariski (1949) reformulated his main theorem in terms of commutative algebra as a statement about local rings. Grothendieck generalized Zariski's formulation as follows:

> If B is an algebra of finite type over a local Noetherian ring A, and n is a maximal ideal of B which is minimal among ideals of B whose inverse image in A is the maximal ideal m of A, then there is a finite A-algebra A' with a maximal ideal m' (whose inverse image in A is m) such that the localization B_n is isomorphic to the A-algebra $A'_{m'}$.

If in addition A and B are integral and have the same field of fractions, and A is integrally closed, then this theorem implies that A and B are equal. This is essentially Zariski's formulation of his main theorem in terms of commutative rings.

Zariski's Main Theorem: Topological Form

A topological version of Zariski's main theorem says that if x is a (closed) point of a normal complex variety it is unibranch; in other words there are arbitrarily small neighborhoods U of x such that the set of non-singular points of U is connected.

The property of being normal is stronger than the property of being unibranch: for example, a cusp of a plane curve is unibranch but not normal.

Zariski's Main Theorem: Power Series Form

A formal power series version of Zariski's main theorem says that if x is a normal point of a variety then it is analytically normal; in other words the completion of the local ring at x is a normal integral domain.

AF+BG Theorem

In algebraic geometry, a field of mathematics, the AF+BG theorem (also known as Max Noether's fundamental theorem) is a result of Max Noether which describes when the equation of an algebraic curve in the complex projective plane can be written in terms of the equations of two other algebraic curves.

Statement

Let F, G, and H be homogeneous polynomials in three variables, with H having higher degree than F and G; let $a = \deg H - \deg F$ and $b = \deg H - \deg G$ (both positive integers) be the differences of the degrees of the polynomials. Suppose that the greatest common divisor of F and G is constant, which means that the projective curves that they define in the projective plane P^2 have an intersection consisting in a finite number of points. For each point P of this intersection, the polynomials F and G generate an ideal $(F, G)_P$ of the local ring of P^2 at P (this local ring is the ring of the fractions n/d, where n and d are polynomials in three variables and $d(P) \neq 0$). The theorem asserts that, if H lies in $(F, G)_P$ for every intersection point P, then there are homogeneous polynomials A and B of degrees a and b, respectively, such that $H = AF + BG$. Furthermore, any two choices of A differ by a multiple of G, and similarly any two choices of B differ by a multiple of F.

Related Results

This theorem may be viewed as a generalization of Bézout's identity, which provides a condition under which an integer or a univariate polynomial h may be expressed as an element of the ideal generated by two other integers or univariate polynomials f and g: such a representation exists exactly when h is a multiple of the greatest common divisor of f and g. The AF+BG condition expresses, in terms of divisors (sets of points, with multiplicities), a similar condition under which a homogeneous polynomial H in three variables can be written as an element of the ideal generated by two other polynomials F and G.

This theorem is also a refinement, for this particular case, of Hilbert's Nullstellensatz, which provides a condition expressing that some power of a polynomial h (in any number of variables) belongs to the ideal generated by a finite set of polynomials.

Grothendieck Trace Formula

In algebraic geometry, the Grothendieck trace formula expresses the number of points of a variety over a finite field in terms of the trace of the Frobenius endomorphism on its cohomology groups. There are several generalizations: the Frobenius endomorphism can be replaced by a more general endomorphism, in which case the points over a finite field are replaced by its fixed points, and there is also a more general version for a sheaf over the variety, where the cohomology groups are replaced by cohomology with coefficients in the sheaf.

The Grothendieck trace formula is an analogue in algebraic geometry of the Lefschetz fixed-point theorem in algebraic topology.

One application of the Grothendieck trace formula is to express the zeta function of a variety over a finite field, or more generally the L-series of a sheaf, as a sum over traces of Frobenius on cohomology groups. This is one of the steps used in the proof of the Weil conjectures.

Behrend's trace formula generalizes the formula to algebraic stacks.

Formal Statement for L-functions

Let k be a finite field, l a prime number invertible in k, X a smooth k-scheme of dimension n, and $\mathcal{F}$ a constructible $\mathbb{Q}_l$-sheaf on X. Then the following cohomological expression for the L-function of $\mathcal{F}$ holds:

$$L(X,\mathcal{F},t)=\prod_{i=0}^{2n}\det(1-t\cdot F\,|\,H_c^i(X_{\bar{k}},\mathcal{F}))^{(-1)^{i+1}}=\frac{\det(1-t\cdot F\,|\,H_c^1(X_{\bar{k}},\mathcal{F}))\ldots\det(1-t\cdot F\,|\,H_c^{2n-1}(X_{\bar{k}},\mathcal{F}))}{\det(1-t\cdot F\,|\,H_c^0(X_{\bar{k}},\mathcal{F}))\ldots\det(1-t\cdot F\,|\,H_c^{2n}(X_{\bar{k}},\mathcal{F}))}$$

where F is everywhere a geometric Frobenius action on l-adic cohomology with compact supports of the sheaf $\mathcal{F}$. Taking logarithmic derivatives of both formal power series produces a statement on sums of traces for each finite field extension E of the base field k:

$$\sum_{x\in X(E)}\operatorname{tr}(F_E\,|\,\mathcal{F}_x)=\sum_{i=0}^{2n}\operatorname{tr}(F_E\,|\,H_c^i(X_{\bar{k}},\mathcal{F}))$$

For a constant sheaf $\mathbb{Q}_l$ (viewed as $(\varprojlim\mathbb{Z}/l^n\mathbb{Z})\otimes\mathbb{Q}$ to qualify as an l-adic sheaf) the left hand side of this formula is the number of E-points of X.

Behrend's Trace Formula

In algebraic geometry, Behrend's formula is a generalization of the Grothendieck–Lefschetz trace formula to a smooth algebraic stack over a finite field, conjectured in 1993 and proven in 2003 by Kai Behrend. Unlike the classical one, the formula counts points in the "stacky way"; it takes into account the presence of nontrivial automorphisms.

The desire for the formula comes from the fact that it applies to the moduli stack of principal bundles on a curve over a finite field (in some instances indirectly, via the Harder–Narasimhan stratification, as the moduli stack is not of finite type.).

Deligne found an example that shows the formula may be interpreted as a sort of the Selberg trace formula.

A proof of the formula in the context of the six operations formalism developed by Laszlo and Olsson is given by Shenghao Sun.

Formulation

By definition, if C is a category in which each object has finitely many automorphisms, the number of points in C is denoted by:

$$\#C = \sum_p \frac{1}{\#\mathrm{Aut}(p)},$$

with the sum running over representatives *p* of all isomorphism classes in *C*. (The series may diverge in general.) The formula states: for a smooth algebraic stack *X* of finite type over a finite field $\mathbf{F}_q$ and the "arithmetic" Frobenius $\phi^{-1}: X \to X$, i.e., the inverse of the usual geometric Frobenius ϕ in Grothendieck's formula,

$$\#X(\mathbf{F}_q) = q^{\dim X} \sum_{i=0}^{\infty} (-1)^i \operatorname{tr}(\phi^{-1}; H^i(X, \mathbb{Q}_l)).$$

Here, it is crucial that the cohomology of a stack is with respect to the smooth topology (not etale).

When *X* is a variety, the smooth cohomology is the same as etale one and, via the Poincaré duality, this is equivalent to Grothendieck's trace formula. (But the proof relies on Grothendieck's formula, so this does not subsume Grothendieck's.)

Simple Example

Consider $B\mathbb{G}_m = [\operatorname{Spec}\mathbf{F}_q / \mathbb{G}_m]$, the classifying stack of the multiplicative group scheme (that is, $\mathbb{G}_m(R) = R^\times$). By definition, $B\mathbb{G}_m(\mathbf{F}_q)$ is the category of principal $\mathbb{G}_m$-bundles over $\operatorname{Spec}\mathbf{F}_q$, which has only one isomorphism class (since all such bundles are trivial by Lang's theorem). Its group of automorphisms is $\mathbb{G}_m$, which means that the number of $\mathbf{F}_q$-isomorphisms is $\#\mathbb{G}_m(\mathbf{F}_q) = \#\mathbf{F}_q^\times = q-1$.

On the other hand, we may compute the *l*-adic cohomology of $B\mathbb{G}_m$ directly. We remark that in the topological setting, we have $B\mathbb{C}^\times \cong \mathbb{CP}^\infty$ (where $B\mathbb{C}^\times$ now denotes the usual classifying space of a topological group), whose rational cohomology ring is a polynomial ring in one generator (Borel's theorem), but we shall not use this directly. If we wish to stay in the world of algebraic geometry, we may instead "approximate" $B\mathbb{G}_m$ by projective spaces of larger and larger dimension. Thus we consider the map $B\mathbb{G}_m \to \mathbb{P}^N$ induced by the $\mathbb{G}_m$-bundle corresponding to $\mathcal{O}(1)$. This map induces an isomorphism in cohomology in degrees up to *2N*. Thus the even (resp. odd) Betti numbers of $B\mathbb{G}_m$ are 1 (resp. 0), and the *l*-adic Galois representation on the *(2n)*th cohomology group is the *n*th power of the cyclotomic character. The second part is a consequence of the fact that the cohomology of $\mathbb{P}^N$ is generated by algebraic cycle classes. This shows that:

$$\sum_{i \geq 0} (-1)^i \operatorname{tr}(\phi^{-1}; H^i(B\mathbb{G}_m, \mathbb{Q}_l)) = 1 + 1/q + 1/q^2 + \cdots = \frac{q}{q-1}.$$

Note that $\dim B\mathbb{G}_m = \dim \operatorname{Spec}\mathbf{F}_q - \dim \mathbb{G}_m = -1$. Multiplying by q^{-1}, one obtains the predicted equality.

Bézout's Theorem

Bézout's theorem is a statement in algebraic geometry concerning the number of common points, or intersection points, of two plane algebraic curves which do not share a common component (that is, which do not have infinitely many common points). The theorem states that the number of common points of two such curves is at most equal to the product of their degrees, and equality holds if one counts points at infinity and points with complex coordinates (or more generally, coordinates from the algebraic closure of the ground field), and if each point is counted with its intersection multiplicity.

Bézout's theorem refers also to the generalization to higher dimensions: Let there be n homogeneous polynomials in n+1 variables, of degrees $d_1,\ldots,d_n$, that define n hypersurfaces in the projective space of dimension n. If the number of intersection points of the hypersurfaces is finite over an algebraic closure of the ground field, then this number is $d_1 \cdots d_n$, if the points are counted with their multiplicity. As in the case of two variables, in the case of affine hypersurfaces, and when not counting multiplicities nor non-real points, this theorem provides only an upper bound of the number of points, which is often reached. This is often referred to as Bézout's bound.

Bézout's theorem is fundamental in computer algebra and effective algebraic geometry, by showing that most problems have a computational complexity that is at least exponential in the number of variables. It follows that in these areas, the best complexity that may be hoped for will occur in algorithms that have a complexity which is polynomial in Bézout's bound.

Rigorous Statement

Suppose that X and Y are two plane projective curves defined over a field F that do not have a common component (this condition means that X and Y are defined by polynomials, whose polynomial greatest common divisor is a constant; in particular, it holds for a pair of "generic" curves). Then the total number of intersection points of X and Y with coordinates in an algebraically closed field E which contains F, counted with their multiplicities, is equal to the product of the degrees of X and Y.

The generalization in higher dimension may be stated as:

Let n projective hypersurfaces be given in a projective space of dimension n over an algebraic closed field, which are defined by n homogeneous polynomials in n + 1 variables, of degrees $d_1,\ldots,d_n$. Then either the number of intersection points is infinite, or the number of intersection points, counted with multiplicity, is equal to the product $d_1 \cdots d_n$. If the hypersurfaces are irreducible and in relative general position, then there are $d_1 \cdots d_n$ intersection points, all with multiplicity 1.

There are various proofs of this theorem. In particular, it may be deduced by applying iteratively the following generalization: if V is a projective algebraic set of dimension δ and degree d_1, and H is a hypersurface (defined by a polynomial) of degree d_2, that does not contain any irreducible component of V, then the intersection of V and H has dimension $\delta-1$ and degree $d_1 d_2$.

Bézout's theorem has been further generalized as the so-called multi-homogeneous Bézout theorem.

Intersection Multiplicity

The most delicate part of Bézout's theorem and its generalization to the case of k algebraic hypersurfaces in k-dimensional projective space is the procedure of assigning the proper intersection multiplicities. If P is a common point of two plane algebraic curves X and Y that is a non-singular point of both of them and, moreover, the tangent lines to X and Y at P are distinct then the intersection multiplicity is one. This corresponds to the case of "transversal intersection". If the curves X and Y have a common tangent at P then the multiplicity is at least two.

Examples

- Two distinct non-parallel lines (in the same plane) always meet in exactly one point. Two parallel lines intersect at a unique point that lies at infinity. To see how this works algebraically, in projective space, the lines $x+2y=3$ and $x+2y=5$ are represented by the homogeneous equations $x+2y-3z=0$ and $x+2y-5z=0$. Solving, we get $x=-2y$ and $z=0$, corresponding to the point (-2:1:0) in homogeneous coordinates. As the z-coordinate is 0, this point lies on the line at infinity.
- The special case where one of the curves is a line can be derived from the fundamental theorem of algebra. In this case the theorem states that an algebraic curve of degree n intersects a given line in n points, counting the multiplicities. For example, the parabola defined by $y-x^2=0$ has degree 2; the line $y-ax=0$ has degree 1, and they meet in exactly two points when $a \neq 0$ and touch at the origin (intersect with multiplicity two) when $a = 0$.
- Two conic sections generally intersect in four points, some of which may coincide. To properly account for all intersection points, it may be necessary to allow complex coordinates and include the points on the infinite line in the projective plane. For example:
    - Two circles never intersect in more than two points in the plane, while Bézout's theorem predicts four. The discrepancy comes from the fact that every circle passes through the same two complex points on the line at infinity. Writing the circle:

  $$(x-a)^2+(y-b)^2=r^2$$

  in homogeneous coordinates, we get:

  $$(x-az)^2+(y-bz)^2-r^2z^2=0$$

  from which it is clear that the two points $(1:i:0)$ and $(1:-i:0)$ lie on every circle. When two circles don't meet at all in the real plane, the two other intersections have non-zero imaginary parts, or if they are concentric then they meet at exactly the two points on the line at infinity with an intersection multiplicity of two.
- Any conic should meet the line at infinity at two points according to the theorem. A hyperbola meets it at two real points corresponding to the two directions of the asymptotes. An ellipse meets it at two complex points which are conjugate to one another---in the case of a circle, the points $(1:i:0)$ and $(1:-i:0)$. A parabola meets it at only one point, but it is a point of tangency and therefore counts twice.
    - The following pictures show examples in which the circle $x^2+y^2-1=0$ meets another ellipse in fewer intersection points because at least one of them has multiplicity greater than 1:

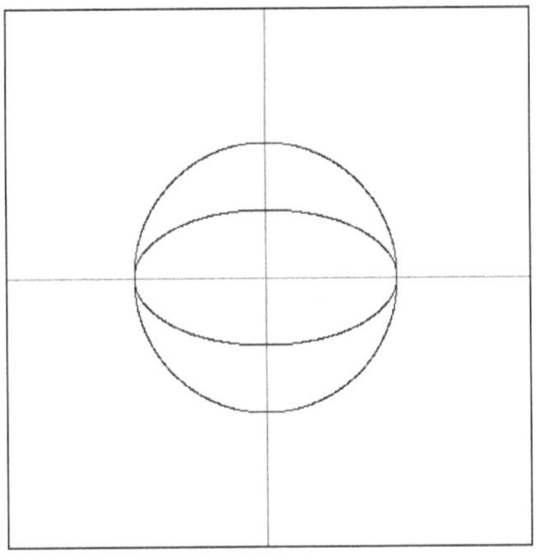

$x^2 + 4y^2 - 1 = 0$: two intersections of multiplicity 2.

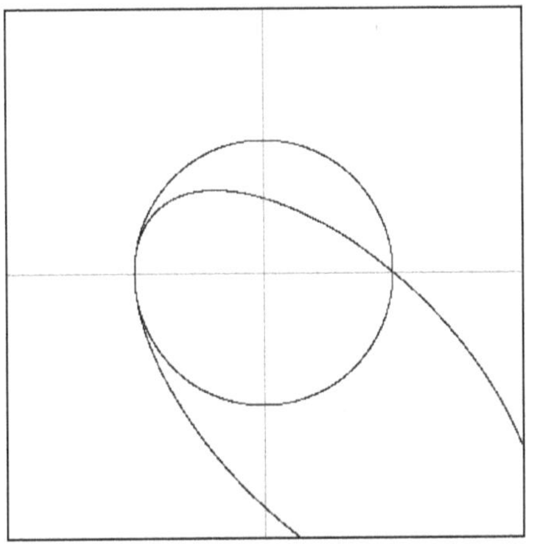

$5x^2 + 6xy + 5y^2 + 6y - 5 = 0$: an intersection of multiplicity 3.

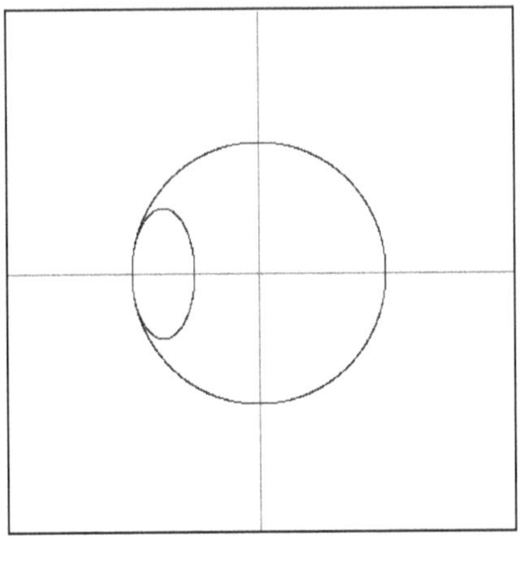

$4x^2 + y^2 + 6x + 2 = 0$: an intersection of multiplicity 4.

Sketch of Proof

Write the equations for *X* and *Y* in homogeneous coordinates as:

$$a_0 z^m + a_1 z^{m-1} + \ldots + a_{m-1} z + a_m = 0$$

$$b_0 z^n + b_1 z^{n-1} + \cdots + b_{n-1} z + b_n = 0$$

where a_i and b_i are homogeneous polynomials of degree i in x and y. The points of intersection of X and Y correspond to the solutions of the system of equations. Form the Sylvester matrix; in the case m=4, n=3 this is:

$$S = \begin{pmatrix} a_0 & a_1 & a_2 & a_3 & a_4 & 0 & 0 \\ 0 & a_0 & a_1 & a_2 & a_3 & a_4 & 0 \\ 0 & 0 & a_0 & a_1 & a_2 & a_3 & a_4 \\ b_0 & b_1 & b_2 & b_3 & 0 & 0 & 0 \\ 0 & b_0 & b_1 & b_2 & b_3 & 0 & 0 \\ 0 & 0 & b_0 & b_1 & b_2 & b_3 & 0 \\ 0 & 0 & 0 & b_0 & b_1 & b_2 & b_3 \end{pmatrix}.$$

The determinant $|S|$ of S, which is also called the resultant of the two polynomials, is 0 exactly when the two equations have a common solution in z. The terms of $|S|$, for example $(a_0)^n(b_n)^m$, all have degree mn, so $|S|$ is a homogeneous polynomial of degree mn in x and y (recall that a_i and b_i are themselves polynomials). By the fundamental theorem of algebra, this can be factored into mn linear factors so there are mn solutions to the system of equations. The linear factors correspond to the lines that join the origin to the points of intersection of the curves.

References

- Tsfasman, Michael A.; Vlăduţ, Serge G.; Nogin, Dmitry. Algebraic Geometric Codes Basic Notions. American Mathematical Soc. ISBN 978-0-8218-7520-9
- Drton, Mathias; Sturmfels, Bernd; Sullivant, Seth (2009). Lectures on Algebraic Statistics. Springer. ISBN 978-3-7643-8904-8
- Shenghao, Sun (2011). "L-series of Artin stacks over finite fields". Algebra & Number Theory. 6: 47–122. arXiv:1008.3689v2. doi:10.2140/ant.2012.6.47
- Falb, Peter. Methods of Algebraic Geometry in Control Theory Part II Multivariable Linear Systems and Projective Algebraic Geometry. Springer. ISBN 978-0-8176-4113-9
- Allen Tannenbaum, Invariance and Systems Theory: Algebraic and Geometric Aspects, Lecture Notes in Mathematics, volume 845, Springer-Verlag, ISBN 9783540105657

6

Theorems Used in Algebra

Fundamental theorems of algebra, Abel-Ruffini theorem, Binomial theorem, Chevalley-Warning theorem, Boolean Prime Ideal theorem, Rational root theorem, factor theorem are the various theorems in algebra. This chapter discusses in detail the theorems related to algebra.

Fundamental Theorem of Algebra

The fundamental theorem of algebra states that every non-constant single-variable polynomial with complex coefficients has at least one complex root. This includes polynomials with real coefficients, since every real number is a complex number with an imaginary part equal to zero.

Equivalently (by definition), the theorem states that the field of complex numbers is algebraically closed.

The theorem is also stated as follows: every non-zero, single-variable, degree n polynomial with complex coefficients has, counted with multiplicity, exactly n roots. The equivalence of the two statements can be proven through the use of successive polynomial division.

In spite of its name, there is no purely algebraic proof of the theorem, since any proof must use the completeness of the reals (or some other equivalent formulation of completeness), which is not an algebraic concept. Additionally, it is not fundamental for modern algebra; its name was given at a time when the study of algebra was mainly concerned with the solutions of polynomial equations with real or complex coefficients.

History

Peter Roth, in his book *Arithmetica Philosophica* (published in 1608, at Nürnberg, by Johann Lantzenberger), wrote that a polynomial equation of degree n (with real coefficients) *may* have n solutions. Albert Girard, in his book *L'invention nouvelle en l'Algèbre* (published in 1629), asserted that a polynomial equation of degree n has n solutions, but he did not state that they had to be real numbers. Furthermore, he added that his assertion holds "unless the equation is incomplete", by which he meant that no coefficient is equal to 0. However, when he explains in detail what he means, it is clear that he actually believes that his assertion is always true; for instance, he shows that the equation $x^4 = 4x - 3$, although incomplete, has four solutions (counting multiplicities): 1 (twice), $-1 + i\sqrt{2}$, and $-1 - i\sqrt{2}$.

As will be mentioned again below, it follows from the fundamental theorem of algebra that every non-constant polynomial with real coefficients can be written as a product of polynomials with real coefficients whose degree is either 1 or 2. However, in 1702 Leibniz said that no polynomial of the type $x^4 + a^4$ (with a real and distinct from 0) can be written in such a way. Later, Nikolaus Bernoulli made the same assertion concerning the polynomial $x^4 - 4x^3 + 2x^2 + 4x + 4$, but he got a letter from Euler in 1742 in which he was told that his polynomial happened to be equal to:

$$(x^2-(2+\alpha)x+1+\sqrt{7}+\alpha)(x^2-(2-\alpha)x+1+\sqrt{7}-\alpha),$$

where α is the square root of $4 + 2\sqrt{7}$. Also, Euler mentioned that:

$$x^4+a^4=(x^2+a\sqrt{2}\cdot x+a^2)(x^2-a\sqrt{2}\cdot x+a^2).$$

A first attempt at proving the theorem was made by d'Alembert in 1746, but his proof was incomplete. Among other problems, it assumed implicitly a theorem (now known as Puiseux's theorem) which would not be proved until more than a century later, and furthermore the proof assumed the fundamental theorem of algebra. Other attempts were made by Euler (1749), de Foncenex (1759), Lagrange (1772), and Laplace (1795). These last four attempts assumed implicitly Girard's assertion; to be more precise, the existence of solutions was assumed and all that remained to be proved was that their form was $a + bi$ for some real numbers a and b. In modern terms, Euler, de Foncenex, Lagrange, and Laplace were assuming the existence of a splitting field of the polynomial $p(z)$.

At the end of the 18th century, two new proofs were published which did not assume the existence of roots, but neither of which was complete. One of them, due to James Wood and mainly algebraic, was published in 1798 and it was totally ignored. Wood's proof had an algebraic gap. The other one was published by Gauss in 1799 and it was mainly geometric, but it had a topological gap, filled by Alexander Ostrowski in 1920, as discussed in Smale 1981 (Smale writes, "...I wish to point out what an immense gap Gauss' proof contained. It is a subtle point even today that a real algebraic plane curve cannot enter a disk without leaving. In fact even though Gauss redid this proof 50 years later, the gap remained. It was not until 1920 that Gauss' proof was completed. In the reference Gauss, A. Ostrowski has a paper which does this and gives an excellent discussion of the problem as well..."). A rigorous proof was first published by Argand in 1806 (and revisited in 1813); it was here that, for the first time, the fundamental theorem of algebra was stated for polynomials with complex coefficients, rather than just real coefficients. Gauss produced two other proofs in 1816 and another version of his original proof in 1849.

The first textbook containing a proof of the theorem was Cauchy's *Cours d'analyse de l'École Royale Polytechnique* (1821). It contained Argand's proof, although Argand is not credited for it.

None of the proofs mentioned so far is constructive. It was Weierstrass who raised for the first time, in the middle of the 19th century, the problem of finding a constructive proof of the fundamental theorem of algebra. He presented his solution, that amounts in modern terms to a combination of the Durand–Kerner method with the homotopy continuation principle, in 1891. Another proof of this kind was obtained by Hellmuth Kneser in 1940 and simplified by his son Martin Kneser in 1981.

Without using countable choice, it is not possible to constructively prove the fundamental theorem of algebra for complex numbers based on the Dedekind real numbers (which are not constructively equivalent to the Cauchy real numbers without countable choice). However, Fred Richman proved a reformulated version of the theorem that does work.

Proofs

All proofs below involve some analysis, or at least the topological concept of continuity of real or complex functions. Some also use differentiable or even analytic functions. This fact has led to the remark that the Fundamental Theorem of Algebra is neither fundamental, nor a theorem of algebra.

Some proofs of the theorem only prove that any non-constant polynomial with real coefficients has some complex root. This is enough to establish the theorem in the general case because, given a non-constant polynomial $p(z)$ with complex coefficients, the polynomial;

$$q(z) = p(z)\overline{p(\overline{z})}$$

has only real coefficients and, if z is a zero of $q(z)$, then either z or its conjugate is a root of $p(z)$.

A large number of non-algebraic proofs of the theorem use the fact (sometimes called "growth lemma") that an n-th degree polynomial function $p(z)$ whose dominant coefficient is 1 behaves like z^n when $|z|$ is large enough. A more precise statement is: there is some positive real number R such that:

$$\tfrac{1}{2}|z^n| < |p(z)| < \tfrac{3}{2}|z^n|$$

when $|z| > R$.

Complex-analytic Proofs

Find a closed disk D of radius r centered at the origin such that $|p(z)| > |p(0)|$ whenever $|z| \geq r$. The minimum of $|p(z)|$ on D, which must exist since D is compact, is therefore achieved at some point z_0 in the interior of D, but not at any point of its boundary. The Maximum modulus principle (applied to $1/p(z)$) implies then that $p(z_0) = 0$. In other words, z_0 is a zero of $p(z)$.

A variation of this proof does not require the use of the maximum modulus principle (in fact, the same argument with minor changes also gives a proof of the maximum modulus principle for holomorphic functions). If we assume by contradiction that $a := p(z_0) \neq 0$, then, expanding $p(z)$ in powers of $z - z_0$ we can write;

$$p(z) = a + c_k(z-z_0)^k + c_{k+1}(z-z_0)^{k+1} + \ldots + c_n(z-z_0)^n.$$

Here, the c_j are simply the coefficients of the polynomial $z \to p(z + z_0)$, and we let k be the index of the first coefficient following the constant term that is non-zero. But now we see that for z sufficiently close to z_0 this has behavior asymptotically similar to the simpler polynomial $q(z) = a + c_k(z-z_0)^k$, in the sense that (as is easy to check) the function $\left|\frac{p(z)-q(z)}{(z-z_0)^{k+1}}\right|$ is bounded by some positive constant M in some neighborhood of z_0. Therefore if we define $\theta_0 = (\arg(a) + \pi - \arg(c_k))/k$ and let $z = z_0 + re^{i\theta_0}$,, then for any sufficiently small positive number r (so that the bound M mentioned above holds), using the triangle inequality we see that;

$$\begin{aligned}|p(z)| &< |q(z)| + r^{k+1}\left|\frac{p(z)-q(z)}{r^{k+1}}\right| \\ &\le \left|a+(-1)c_k r^k e^{i(\arg(a)-\arg(c_k))}\right| + Mr^{k+1} \\ &= |a| - |c_k| r^k + Mr^{k+1}.\end{aligned}$$

When r is sufficiently close to 0 this upper bound for $|p(z)|$ is strictly smaller than $|a|$, in contradiction to the definition of z_0. (Geometrically, we have found an explicit direction θ_0 such that if one approaches z_0 from that direction one can obtain values $p(z)$ smaller in absolute value than $|p(z_0)|$.)

Another analytic proof can be obtained along this line of thought observing that, since $|p(z)| > |p(0)|$ outside D, the minimum of $|p(z)|$ on the whole complex plane is achieved at z_0. If $|p(z_0)| > 0$, then $1/p$ is a bounded holomorphic function in the entire complex plane since, for each complex number z, $|1/p(z)| \le |1/p(z_0)|$. Applying Liouville's theorem, which states that a bounded entire function must be constant, this would imply that $1/p$ is constant and therefore that p is constant. This gives a contradiction, and hence $p(z_0) = 0$.

Yet another analytic proof uses the argument principle. Let R be a positive real number large enough so that every root of $p(z)$ has absolute value smaller than R; such a number must exist because every non-constant polynomial function of degree n has at most n zeros. For each $r > R$, consider the number;

$$\frac{1}{2\pi i}\int_{c(r)} \frac{p'(z)}{p(z)}dz,$$

where $c(r)$ is the circle centered at 0 with radius r oriented counterclockwise; then the argument principle says that this number is the number N of zeros of $p(z)$ in the open ball centered at 0 with radius r, which, since $r > R$, is the total number of zeros of $p(z)$. On the other hand, the integral of n/z along $c(r)$ divided by $2\pi i$ is equal to n. But the difference between the two numbers is;

$$\frac{1}{2\pi i}\int_{c(r)}\left(\frac{p'(z)}{p(z)}-\frac{n}{z}\right)dz = \frac{1}{2\pi i}\int_{c(r)}\frac{zp'(z)-np(z)}{zp(z)}dz.$$

The numerator of the rational expression being integrated has degree at most $n - 1$ and the degree of the denominator is $n + 1$. Therefore, the number above tends to 0 as $r \to +\infty$. But the number is also equal to $N - n$ and so $N = n$.

Still another complex-analytic proof can be given by combining linear algebra with the Cauchy theorem. To establish that every complex polynomial of degree $n > 0$ has a zero, it suffices to show that every complex square matrix of size $n > 0$ has a (complex) eigenvalue. The proof of the latter statement is by contradiction.

Let A be a complex square matrix of size $n > 0$ and let I_n be the unit matrix of the same size. Assume A has no eigenvalues. Consider the resolvent function;

$$R(z) = (zI_n - A)^{-1},$$

which is a meromorphic function on the complex plane with values in the vector space of matrices. The eigenvalues of A are precisely the poles of $R(z)$. Since, by assumption, A has no eigenvalues, the function $R(z)$ is an entire function and Cauchy theorem implies that;

$$\int_{c(r)} R(z)dz = 0.$$

On the other hand, $R(z)$ expanded as a geometric series gives:

$$R(z) = z^{-1}(I_n - z^{-1}A)^{-1} = z^{-1}\sum_{k=0}^{\infty}\frac{1}{z^k}A^k.$$

This formula is valid outside the closed disc of radius $||A||$ (the operator norm of A). Let $r > ||A||$. Then

$$\int_{c(r)} R(z)dz = \sum_{k=0}^{\infty}\int_{c(r)}\frac{dz}{z^{k+1}}A^k = 2\pi i I_n$$

(in which only the summand $k = 0$ has a nonzero integral). This is a contradiction, and so A has an eigenvalue.

Finally, Rouché's theorem gives perhaps the shortest proof of the theorem.

Topological Proofs

Let $z_0 \in \mathbb{C}$ be such that the minimum of $|p(z)|$ on the whole complex plane is achieved at z_0; it was seen at the proof which uses Liouville's theorem that such a number must exist. We can write $p(z)$ as a polynomial in $z - z_0$: there is some natural number k and there are some complex numbers c_k, c_{k+1}, ..., c_n such that $c_k \neq 0$ and that;

$$p(z) = p(z_0) + c_k(z - z_0)^k + c_{k+1}(z - z_0)^{k+1} + \cdots + c_n(z - z_0)^n.$$

In the case that $p(z_0)$ is nonzero, it follows that if a is a kth root of $-p(z_0)/c_k$ and if t is positive and sufficiently small, then $|p(z_0 + ta)| < |p(z_0)|$, which is impossible, since $|p(z_0)|$ is the minimum of $|p|$ on D.

For another topological proof by contradiction, suppose that $p(z)$ has no zeros. Choose a large positive number R such that, for $|z| = R$, the leading term z^n of $p(z)$ dominates all other terms combined; in other words, such that $|z|^n > |a_{n-1}z^{n-1} + \cdots + a_0|$. As z traverses the circle given by the equation $|z| = R$ once counter-clockwise, $p(z)$, like z^n, winds n times counter-clockwise around 0. At the other extreme, with $|z| = 0$, the "curve" $p(z)$ is simply the single (nonzero) point $p(0)$, whose winding number is clearly 0. If the loop followed by z is continuously deformed between these extremes, the path of $p(z)$ also deforms continuously. We can explicitly write such a deformation as $H(Re^{i\theta},t) = p((1 - t)Re^{i\theta})$, where $0 \leq t \leq 1$. If one views the variable t as time, then at time zero the curve is $p(z)$ and at time one the curve is $p(0)$. Clearly at every point t, $p(z)$ cannot be zero by the original assumption, therefore during the deformation, the curve never crosses zero. Therefore the winding number of the curve around zero should never change. However, given that the winding number started as n and ended as 0, this is absurd. Therefore, $p(z)$ has at least one zero.

Algebraic Proofs

These proofs use two facts about real numbers that require only a small amount of analysis (more precisely, the intermediate value theorem):

- every polynomial with odd degree and real coefficients has some real root;
- every non-negative real number has a square root.

The second fact, together with the quadratic formula, implies the theorem for real quadratic polynomials. In other words, algebraic proofs of the fundamental theorem actually show that if R is any real-closed field, then its extension $C = R(\sqrt{-1})$ is algebraically closed.

As mentioned above, it suffices to check the statement "every non-constant polynomial $p(z)$ with real coefficients has a complex root". This statement can be proved by induction on the greatest non-negative integer k such that 2^k divides the degree n of $p(z)$. Let a be the coefficient of z^n in $p(z)$ and let F be a splitting field of $p(z)$ over C; in other words, the field F contains C and there are elements $z_1, z_2, \ldots, z_n$ in F such that;

$$p(z) = a(z - z_1)(z - z_2)\cdots(z - z_n).$$

If $k = 0$, then n is odd, and therefore $p(z)$ has a real root. Now, suppose that $n = 2^k m$ (with m odd and $k > 0$) and that the theorem is already proved when the degree of the polynomial has the form $2^{k-1}m'$ with m' odd. For a real number t, define:

$$q_t(z) = \prod_{1 \le i < j \le n} \left(z - z_i - z_j - t z_i z_j\right).$$

Then the coefficients of $q_t(z)$ are symmetric polynomials in the z_i's with real coefficients. Therefore, they can be expressed as polynomials with real coefficients in the elementary symmetric polynomials, that is, in $-a_1, a_2, \ldots, (-1)^n a_n$. So $q_t(z)$ has in fact *real* coefficients. Furthermore, the degree of $q_t(z)$ is $n(n-1)/2 = 2^{k-1}m(n-1)$, and $m(n-1)$ is an odd number. So, using the induction hypothesis, q_t has at least one complex root; in other words, $z_i + z_j + t z_i z_j$ is complex for two distinct elements i and j from $\{1, \ldots, n\}$. Since there are more real numbers than pairs (i, j), one can find distinct real numbers t and s such that $z_i + z_j + t z_i z_j$ and $z_i + z_j + s z_i z_j$ are complex (for the same i and j). So, both $z_i + z_j$ and $z_i z_j$ are complex numbers. It is easy to check that every complex number has a complex square root, thus every complex polynomial of degree 2 has a complex root by the quadratic formula. It follows that z_i and z_j are complex numbers, since they are roots of the quadratic polynomial $z^2 - (z_i + z_j)z + z_i z_j$.

Joseph Shipman showed in 2007 that the assumption that odd degree polynomials have roots is stronger than necessary; any field in which polynomials of prime degree have roots is algebraically closed (so "odd" can be replaced by "odd prime" and furthermore this holds for fields of all characteristics). For axiomatization of algebraically closed fields, this is the best possible, as there are counterexamples if a single prime is excluded. However, these counterexamples rely on -1 having a square root. If we take a field where -1 has no square root, and every polynomial of degree $n \in I$ has a root, where I is any fixed infinite set of odd numbers, then every polynomial $f(x)$ of odd degree has a root (since $(x^2 + 1)^k f(x)$ has a root, where k is chosen so that $\deg(f) + 2k \in I$). Mohsen Aliabadi generalized Shipman's result for any field in 2013, proving that the sufficient condition for an arbitrary field (of any characteristic) to be algebraically closed is having a root for any polynomial of prime degree.

Another algebraic proof of the fundamental theorem can be given using Galois theory. It suffices to show that C has no proper finite field extension. Let K/C be a finite extension. Since the normal closure of K over R still has a finite degree over C (or R), we may assume without loss of generality that K is a normal extension of R (hence it is a Galois extension, as every algebraic extension of a field of characteristic 0 is separable). Let G be the Galois group of this extension, and let H be a Sylow 2-subgroup of G, so that the order of H is a power of 2, and the index of H in G is odd. By the fundamental theorem of Galois theory, there exists a subextension L of K/R such that $\mathrm{Gal}(K/L) = H$. As $[L:R] = [G:H]$ is odd, and there are no nonlinear irreducible real polynomials of odd degree, we must have $L = R$, thus $[K:R]$ and $[K:C]$ are powers of 2. Assuming by way of contradiction that $[K:C] > 1$, we conclude that the 2-group $\mathrm{Gal}(K/C)$ contains a subgroup of index 2, so there exists a subextension M of C of degree 2. However, C has no extension of degree 2, because every quadratic complex polynomial has a complex root, as mentioned above. This shows that $[K:C] = 1$, and therefore $K = C$, which completes the proof.

Geometric Proofs

There exists still another way to approach the fundamental theorem of algebra, due to J. M. Almira and A. Romero: by Riemannian Geometric arguments. The main idea here is to prove that the existence of a non-constant polynomial $p(z)$ without zeros implies the existence of a flat Riemannian metric over the sphere S^2. This leads to a contradiction, since the sphere is not flat.

Recall that a Riemannian surface (M, g) is said to be flat if its Gaussian curvature, which we denote by K_g, is identically null. Now, Gauss–Bonnet theorem, when applied to the sphere S^2, claims that;

$$\int_{S^2} K_g = 4\pi ,$$

which proves that the sphere is not flat.

Let us now assume that $n > 0$ and $p(z) = a_0 + a_1 z + \cdots + a_n z^n \neq 0$ for each complex number z. Let us define $p^*(z) = z^n p(1/z) = a_0 z^n + a_1 z^{n-1} + \cdots + a_n$. Obviously, $p^*(z) \neq 0$ for all z in C. Consider the polynomial $f(z) = p(z)p^*(z)$. Then $f(z) \neq 0$ for each z in C. Furthermore,

$$f(\tfrac{1}{w}) = p(\tfrac{1}{w})p^*(\tfrac{1}{w}) = w^{-2n} p^*(w)p(w) = w^{-2n} f(w).$$

We can use this functional equation to prove that g, given by;

$$g = \frac{1}{|f(w)|^{\frac{2}{n}}} |dw|^2$$

for w in C, and

$$g = \frac{1}{|f(1/w)|^{\frac{2}{n}}} |d(1/w)|^2$$

for $w \in S^2\setminus\{0\}$, is a well defined Riemannian metric over the sphere S^2 (which we identify with the extended complex plane $C \cup \{\infty\}$).

Now, a simple computation shows that;

$$\forall w \in \mathbf{C}: \frac{1}{|f(w)|^{\frac{1}{n}}} K_g = \frac{1}{n}\Delta \log|f(w)| = \frac{1}{n}\Delta \mathrm{Re}(\log f(w)) = 0,$$

since the real part of an analytic function is harmonic. This proves that $K_g = 0$.

Corollaries

Since the fundamental theorem of algebra can be seen as the statement that the field of complex numbers is algebraically closed, it follows that any theorem concerning algebraically closed fields applies to the field of complex numbers. Here are a few more consequences of the theorem, which are either about the field of real numbers or about the relationship between the field of real numbers and the field of complex numbers:

- The field of complex numbers is the algebraic closure of the field of real numbers.
- Every polynomial in one variable z with complex coefficients is the product of a complex constant and polynomials of the form $z + a$ with a complex.
- Every polynomial in one variable x with real coefficients can be uniquely written as the product of a constant, polynomials of the form $x + a$ with a real, and polynomials of the form $x^2 + ax + b$ with a and b real and $a^2 - 4b < 0$ (which is the same thing as saying that the polynomial $x^2 + ax + b$ has no real roots). This implies that the number of non-real complex roots (up to multiplicity) is always even.
- Every rational function in one variable x, with real coefficients, can be written as the sum of a polynomial function with rational functions of the form $a/(x - b)^n$ (where n is a natural number, and a and b are real numbers), and rational functions of the form $(ax + b)/(x^2 + cx + d)^n$ (where n is a natural number, and a, b, c, and d are real numbers such that $c^2 - 4d < 0$). A corollary of this is that every rational function in one variable and real coefficients has an elementary primitive.
- Every algebraic extension of the real field is isomorphic either to the real field or to the complex field.

Abel–Ruffini Theorem

In algebra, the Abel–Ruffini theorem (also known as Abel's impossibility theorem) states that there is no algebraic solution—that is, solution in radicals—to the general polynomial equations of degree five or higher with arbitrary coefficients. The theorem is named after Paolo Ruffini, who made an incomplete proof in 1799, and Niels Henrik Abel, who provided a proof in 1824.

Interpretation

The theorem does *not* assert that some higher-degree polynomial equations have *no* solution. In fact, the opposite is true: *every* non-constant polynomial equation in one unknown, with real or complex coefficients, has at least one complex number as a solution (and thus, by polynomial division, as many complex roots as its degree, counting repeated roots); this is the fundamental theorem of algebra. These solutions can be computed to any desired degree of accuracy using numerical methods such as the Newton–Raphson method or the Laguerre method, and in this way they are no different from solutions to polynomial equations of the second, third, or fourth degrees. The theorem only shows that there is no *general* solution in radicals that applies to *all* equations of a given degree greater than 4.

The solution of any second-degree polynomial equation can be expressed in terms of its coefficients, using only addition, subtraction, multiplication, division, and square roots, in the familiar quadratic formula: the roots of the equation $ax^2 + bx + c = 0$ (with $a \neq 0$) are;

$$\frac{-b \pm \sqrt{b^2 - 4ac}}{2a}.$$

Analogous formulas for third-degree equations and fourth-degree equations (using square roots and cube roots) have been known since the 16th century. What the Abel–Ruffini theorem says is that there is no similar formula for general equations of fifth degree or higher. In principle, it could be that the equations of the fifth degree could be split in several types and, for each one of these types, there could be some algebraic solution valid within that type. Or, as Ian Stewart wrote, "for all that Abel's methods could prove, every particular quintic equation might be soluble, with a special formula for each equation." However, this is not so, but this impossibility lies outside the scope of the Abel–Ruffini theorem and is part of the Galois theory.

Proof

The following proof is based on Galois theory and it is valid for any field of characteristic 0. Historically, Ruffini and Abel's proofs precede Galois theory. For a modern presentation of Abel's proof see the article of Rosen or the books of Tignol or Pesic.

One of the fundamental theorems of Galois theory states that a polynomial $P(x) \in F[x]$ is solvable by radicals over F if and only if its splitting field K over F has a solvable Galois group, so the proof of the Abel–Ruffini theorem comes down to computing the Galois group of the general polynomial of the fifth degree.

Consider five indeterminates y_1, y_2, y_3, y_4, and y_5, let $E = \mathbb{Q}(y_1, y_2, y_3, y_4, y_5)$, and let

$$P(x) = (x - y_1)(x - y_2)(x - y_3)(x - y_4)(x - y_5) \in E[x].$$

Expanding $P(x)$ out yields the elementary symmetric functions of the y_i:

$$s_1 = y_1 + y_2 + y_3 + y_4 + y_5,$$

$$s_2 = y_1y_2 + y_1y_3 + y_1y_4 + y_1y_5 + y_2y_3 + y_2y_4 + y_2y_5 + y_3y_4 + y_3y_5 + y_4y_5,$$

$s_3 = y_1y_2y_3 + y_1y_2y_4 + y_1y_2y_5 + y_1y_3y_4 + y_1y_3y_5 + y_1y_4y_5 + y_2y_3y_4 + y_2y_3y_5 + y_2y_4y_5 + y_3y_4y_5,$

$s_4 = y_1y_2y_3y_4 + y_1y_2y_3y_5 + y_1y_2y_4y_5 + y_1y_3y_4y_5 + y_2y_3y_4y_5,$

$s_5 = y_1y_2y_3y_4y_5.$

The coefficient of x^n in $P(x)$ is thus $(-1)^{5-n}s_{5-n}$. Let $F = Q(s_1, s_2, s_3, s_4, s_5)$ be the field obtained by adjoining the symmetric functions to the rationals. Then $P(x) \in F[x]$. Because the y_i's are indeterminates, every permutation σ in the symmetric group on 5 letters S_5 induces a distinct automorphism σ' on E that leaves Q fixed and permutes the elements y_i. Since an arbitrary rearrangement of the roots of the product form still produces the same polynomial, e.g.

$$(x - y_3)(x - y_1)(x - y_2)(x - y_5)(x - y_4)$$

is the same polynomial as

$$(x - y_1)(x - y_2)(x - y_3)(x - y_4)(x - y_5),$$

the automorphisms σ' also leave F fixed, so they are elements of the Galois group Gal(E/F). Therefore, we have shown that $S_5 \subseteq$ Gal(E/F); however there could possibly be automorphisms there that are not in S_5. But, since the Galois group of the splitting field of a quintic polynomial has at most 5! elements, and since E is a splitting field of $P(x)$, it follows that Gal(E/F) is isomorphic to S_5. Generalizing this argument shows that the Galois group of every general polynomial of degree n is isomorphic to S_n.

And what of S_5? The only composition series of S_5 is $S_5 \geq A_5 \geq \{e\}$ (where A_5 is the alternating group on five letters, also known as the icosahedral group). However, the quotient group $A_5/\{e\}$ (isomorphic to A_5 itself) is not abelian, and so S_5 is not solvable, so it must be that the general polynomial of the fifth degree has no solution in radicals. Since the first nontrivial normal subgroup of the symmetric group on n letters is always the alternating group on n letters, and since the alternating groups on n letters for $n \geq 5$ are always simple and non-abelian, and hence not solvable, it also says that the general polynomials of all degrees higher than the fifth also have no solution in radicals. Q.E.D.

The above construction of the Galois group for a fifth degree polynomial only applies to the *general polynomial*; specific polynomials of the fifth degree may have different Galois groups with quite different properties, e.g. $x^5 - 1$ has a splitting field generated by a primitive 5th root of unity, and hence its Galois group is abelian and the equation itself solvable by radicals; moreover, the argument does not provide any rational-valued quintic that has S_5 or A_5 as its Galois group. However, since the result is on the general polynomial, it does say that a general "quintic formula" for the roots of a quintic using only a finite combination of the arithmetic operations and radicals in terms of the coefficients is impossible.v

The proof is not valid if applied to polynomials whose degree is less than 5. Indeed:v

- the group A_4 is *not* simple, because the subgroup {e, (12)(34), (13)(24), (14)(23)} is a normal subgroup;
- the groups A_2 and A_3 *are* simple, but since they are abelian too (A_2 is the trivial group and A_3 is the cyclic group of order 3), that is not a problem.

The proof remains valid if, instead of working with five indeterminates, one works with five concrete algebraically independent complex numbers, because, by the same argument, Gal(E/F) = S_5.

History

Around 1770, Joseph Louis Lagrange began the groundwork that unified the many different tricks that had been used up to that point to solve equations, relating them to the theory of groups of permutations, in the form of Lagrange resolvents. This innovative work by Lagrange was a precursor to Galois theory, and its failure to develop solutions for equations of fifth and higher degrees hinted that such solutions might be impossible, but it did not provide conclusive proof. The first person who conjectured that the problem of solving quintics by radicals might be impossible to solve was Carl Friedrich Gauss, who wrote in 1798 in section 359 of his book *Disquisitiones Arithmeticae* (which would be published only in 1801) that "there is little doubt that this problem does not so much defy modern methods of analysis as that it proposes the impossible". The next year, in his thesis, he wrote "After the labors of many geometers left little hope of ever arriving at the resolution of the general equation algebraically, it appears more and more likely that this resolution is impossible and contradictory." And he added "Perhaps it will not be so difficult to prove, with all rigor, the impossibility for the fifth degree. I shall set forth my investigations of this at greater length in another place." Actually, Gauss published nothing else on this subject.

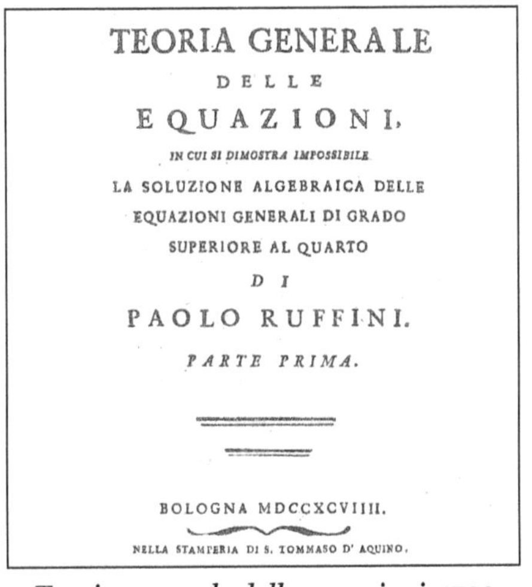
TEORIA GENERALE
DELLE
EQUAZIONI,
IN CUI SI DIMOSTRA IMPOSSIBILE
LA SOLUZIONE ALGEBRAICA DELLE
EQUAZIONI GENERALI DI GRADO
SUPERIORE AL QUARTO
DI
PAOLO RUFFINI.
PARTE PRIMA.

BOLOGNA MDCCXCVIIII.
NELLA STAMPERIA DI S. TOMMASO D' AQUINO.

Teoria generale delle equazioni, 1799.

The theorem was first nearly proved by Paolo Ruffini in 1799. He sent his proof to several mathematicians to get it acknowledged, amongst them Lagrange (who did not reply) and Augustin-Louis Cauchy, who sent him a letter saying: "Your memoir on the general solution of equations is a work which I have always believed should be kept in mind by mathematicians and which, in my opinion, proves conclusively the algebraic unsolvability of general equations of higher than fourth degree." However, in general, Ruffini's proof was not considered convincing. Abel wrote: "The first and, if I am not mistaken, the only one who, before me, has sought to prove the impossibility of the algebraic solution of general equations is the mathematician Ruffini. But his memoir is so complicated that it is very difficult to determine the validity of his argument. It seems to me that his argument is not completely satisfying."

The proof also, as it was discovered later, was incomplete. Ruffini assumed that all radicals that he was dealing with could be expressed from the roots of the polynomial using field operations alone; in modern terms, he assumed that the radicals belonged to the splitting field of the polynomial. To see why this is really an extra assumption, consider, for instance, the polynomial $P(x) = x^3 - 15x - 20$. According to Cardano's formula, one of its roots (all of them, actually) can be expressed as the sum of a cube root of 10 + 5i with a cube root of 10 − 5i. On the other hand, since $P(-3) < 0$, $P(-2) > 0$, $P(-1) < 0$, and $P(5) > 0$, the roots r_1, r_2, and r_3 of $P(x)$ are all real and therefore the field $Q(r_1, r_2, r_3)$ is a subfield of R. But then the numbers $10 \pm 5i$ cannot belong to $Q(r_1, r_2, r_3)$. While Cauchy either did not notice Ruffini's assumption or felt that it was a minor one, most historians believe that the proof was not complete until Abel proved the theorem on natural irrationalities, which asserts that the assumption holds in the case of general polynomials. The Abel-Ruffini theorem is thus generally credited to Abel, who published a proof in just six pages in 1824. However, this short number of pages was obtained at the cost of writing in a very terse style. This was due to the fact that he had the proof printed at his own expenses and he needed to save paper and money. A more elaborated version of the proof would be published in 1826.

Proving that the general quintic (and higher) equations were unsolvable by radicals did not completely settle the matter, because the Abel–Ruffini theorem does not provide necessary and sufficient conditions for saying precisely which quintic (and higher) equations are unsolvable by radicals. Abel was working on a complete characterization when he died in 1829.

According to Nathan Jacobson, "The proofs of Ruffini and of Abel [...] were soon superseded by the crowning achievement of this line of research: Galois' discoveries in the theory of equations." In 1830, Galois (at the age of 18) submitted to the Paris Academy of Sciences a memoir on his theory of solvability by radicals, which was ultimately rejected in 1831 as being too sketchy and for giving a condition in terms of the roots of the equation instead of its coefficients. Galois was aware of the contributions of Ruffini and Abel, since he wrote "It is a common truth, today, that the general equation of degree greater than 4 cannot be solved by radicals... this truth has become common (by hearsay) despite the fact that geometers have ignored the proofs of Abel and Ruffini..." Galois then died in 1832 and his paper *Mémoire sur les conditions de resolubilité des équations par radicaux* remained unpublished until 1846, when it was published by Joseph Liouville accompanied by some of his own explanations. Prior to this publication, Liouville announced Galois' result to the Academy in a speech he gave on 4 July 1843. A simplification of Abel's proof was published by Pierre Wantzel in 1845. When he published it, he was already aware of the contributions by Galois and he mentions that, whereas Abel's proof is valid only for general polynomials, Galois' approach can be used to provide a concrete polynomial of degree 5 whose roots cannot be expressed in radicals from its coefficients.

In 1963, Vladimir Arnold discovered a topological proof of the Abel–Ruffini theorem, which served as a starting point for topological Galois theory.

Binomial Theorem

1
1 1
1 2 1
1 3 3 1
1 4 6 4 1
1 5 10 10 5 1

The binomial coefficients appear as the entries of Pascal's triangle where each entry is the sum of the two above it.

In elementary algebra, the binomial theorem (or binomial expansion) describes the algebraic expansion of powers of a binomial. According to the theorem, it is possible to expand the power $(x + y)^n$ into a sum involving terms of the form ax^by^c, where the exponents b and c are nonnegative integers with $b + c = n$, and the coefficient a of each term is a specific positive integer depending on n and b. For example,

$$(x+y)^4 = x^4 + 4x^3y + 6x^2y^2 + 4xy^3 + y^4.$$

The coefficient a in the term of ax^by^c is known as the binomial coefficient $\binom{n}{b}$ or $\binom{n}{c}$ (the two have the same value). These coefficients for varying n and b can be arranged to form Pascal's triangle. These numbers also arise in combinatorics, where $\binom{n}{b}$ gives the number of different combinations of b elements that can be chosen from an n-element set.

History

Special cases of the binomial theorem were known from ancient times. The 4th century B.C. Greek mathematician Euclid mentioned the special case of the binomial theorem for exponent 2. There is evidence that the binomial theorem for cubes was known by the 6th century in India.

Binomial coefficients, as combinatorial quantities expressing the number of ways of selecting k objects out of n without replacement, were of interest to the ancient Hindus. The earliest known reference to this combinatorial problem is the *Chanda śāstra* by the Hindu lyricist Pingala (c. 200 B.C.), which contains a method for its solution. The commentator Halayudha from the 10th century A.D. explains this method using what is now known as Pascal's triangle. By the 6th century A.D., the Hindu mathematicians probably knew how to express this as a quotient $\frac{n!}{(n-k)!k!}$, and a clear statement of this rule can be found in the 12th century text *Lilavati* by Bhaskara.

The binomial theorem as such can be found in the work of 11th-century Persian mathematician Al-Karaji, who described the triangular pattern of the binomial coefficients. He also provided a mathematical proof of both the binomial theorem and Pascal's triangle, using a primitive form of mathematical induction. The Persian poet and mathematician Omar Khayyam was probably familiar with the formula to higher orders, although many of his mathematical works are lost. The binomial expansions of small degrees were known in the 13th century mathematical works of Yang Hui and also Chu Shih-Chieh. Yang Hui attributes the method to a much earlier 11th century text of Jia Xian, although those writings are now also lost.

In 1544, Michael Stifel introduced the term "binomial coefficient" and showed how to use them to express $(1+a)^n$ in terms of $(1+a)^{n-1}$, via "Pascal's triangle". Blaise Pascal studied the eponymous triangle comprehensively in the treatise *Traité du triangle arithmétique* (1653). However, the pattern of numbers was already known to the European mathematicians of the late Renaissance, including Stifel, Niccolò Fontana Tartaglia, and Simon Stevin.

Isaac Newton is generally credited with the generalized binomial theorem, valid for any rational exponent.

Theorem Statement

According to the theorem, it is possible to expand any power of $x + y$ into a sum of the form:

$$(x+y)^n = \binom{n}{0}x^n y^0 + \binom{n}{1}x^{n-1}y^1 + \binom{n}{2}x^{n-2}y^2 + \cdots + \binom{n}{n-1}x^1 y^{n-1} + \binom{n}{n}x^0 y^n,$$

where each $\binom{n}{k}$ is a specific positive integer known as a binomial coefficient. (When an exponent is zero, the corresponding power expression is taken to be 1 and this multiplicative factor is often omitted from the term. Hence one often sees the right side written as $\binom{n}{0}x^n + \ldots$).) This formula is also referred to as the binomial formula or the binomial identity. Using summation notation, it can be written as:

$$(x+y)^n = \sum_{k=0}^{n}\binom{n}{k}x^{n-k}y^k = \sum_{k=0}^{n}\binom{n}{k}x^k y^{n-k}.$$

The final expression follows from the previous one by the symmetry of x and y in the first expression, and by comparison it follows that the sequence of binomial coefficients in the formula is symmetrical. A simple variant of the binomial formula is obtained by substituting 1 for y, so that it involves only a single variable. In this form, the formula reads:

$$(1+x)^n = \binom{n}{0}x^0 + \binom{n}{1}x^1 + \binom{n}{2}x^2 + \cdots + \binom{n}{n-1}x^{n-1} + \binom{n}{n}x^n,$$

or equivalently

$$(1+x)^n = \sum_{k=0}^{n}\binom{n}{k}x^k.$$

Examples

							1							
						1		1						
					1		2		1					
				1		3		3		1				
			1		4		6		4		1			
		1		5		10		10		5		1		
	1		6		15		20		15		6		1	
1		7		21		35		35		21		7		1

Pascal's triangle.

The most basic example of the binomial theorem is the formula for the square of $x + y$:

$$(x+y)^2 = x^2 + 2xy + y^2.$$

The binomial coefficients 1, 2, 1 appearing in this expansion correspond to the second row of Pascal's triangle. (Note that the top "1" of the triangle is considered to be row 0, by convention.) The coefficients of higher powers of $x + y$ correspond to lower rows of the triangle:

$$(x+y)^3 = x^3 + 3x^2y + 3xy^2 + y^3,$$
$$(x+y)^4 = x^4 + 4x^3y + 6x^2y^2 + 4xy^3 + y^4,$$
$$(x+y)^5 = x^5 + 5x^4y + 10x^3y^2 + 10x^2y^3 + 5xy^4 + y^5,$$
$$(x+y)^6 = x^6 + 6x^5y + 15x^4y^2 + 20x^3y^3 + 15x^2y^4 + 6xy^5 + y^6,$$
$$(x+y)^7 = x^7 + 7x^6y + 21x^5y^2 + 35x^4y^3 + 35x^3y^4 + 21x^2y^5 + 7xy^6 + y^7.$$

Several patterns can be observed from these examples. In general, for the expansion $(x + y)^n$:

1. the powers of x start at n and decrease by 1 in each term until they reach 0 (with $x^0 = 1$ often unwritten);
2. the powers of y start at 0 and increase by 1 until they reach n;
3. the nth row of Pascal's Triangle will be the coefficients of the expanded binomial when the terms are arranged in this way;
4. the number of terms in the expansion before like terms are combined is the sum of the coefficients and is equal to 2^n;
5. there will be $n + 1$ terms in the expression after combining like terms in the expansion.

The binomial theorem can be applied to the powers of any binomial. For example,

$$\begin{aligned}(x+2)^3 &= x^3 + 3x^2(2) + 3x(2)^2 + 2^3 \\ &= x^3 + 6x^2 + 12x + 8.\end{aligned}$$

For a binomial involving subtraction, the theorem can be applied by using the form $(x - y)^n = (x + (-y))^n$. This has the effect of changing the sign of every other term in the expansion:

$$(x-y)^3 = (x+(-y))^3 = x^3 + 3x^2(-y) + 3x(-y)^2 + (-y)^3 = x^3 - 3x^2y + 3xy^2 - y^3.$$

Geometric Explanation

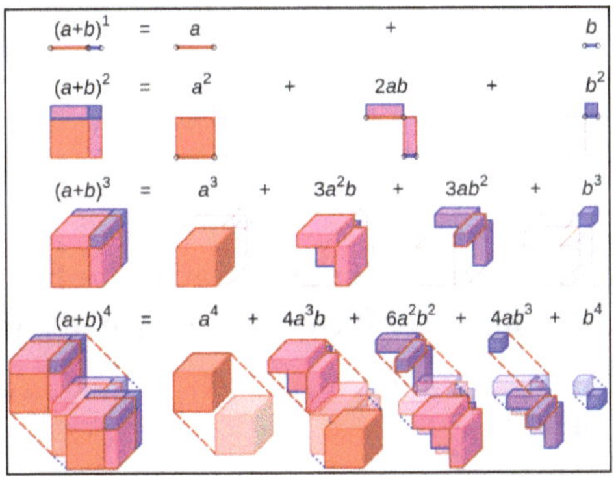

Visualisation of binomial expansion up to the 4th power.

For positive values of a and b, the binomial theorem with $n = 2$ is the geometrically evident fact that a square of side $a + b$ can be cut into a square of side a, a square of side b, and two rectangles with sides a and b. With $n = 3$, the theorem states that a cube of side $a + b$ can be cut into a cube of side a, a cube of side b, three $a{\times}a{\times}b$ rectangular boxes, and three $a{\times}b{\times}b$ rectangular boxes.

In calculus, this picture also gives a geometric proof of the derivative $(x^n)' = nx^{n-1}$: if one sets $a = x$ and $b = \Delta x$, interpreting b as an infinitesimal change in a, then this picture shows the infinitesimal change in the volume of an n-dimensional hypercube, $(x+\Delta x)^n$, where the coefficient of the linear term (in Δx) is nx^{n-1}, the area of the n faces, each of dimension $(n-1)$:

$$(x+\Delta x)^n = x^n + nx^{n-1}\Delta x + \tbinom{n}{2}x^{n-2}(\Delta x)^2 + \cdots.$$

Substituting this into the definition of the derivative via a difference quotient and taking limits means that the higher order terms – $(\Delta x)^2$ and higher – become negligible, and yields the formula $(x^n)' = nx^{n-1}$, interpreted as:

> "the infinitesimal rate of change in volume of an n-cube as side length varies is the area of n of its $(n-1)$-dimensional faces".

If one integrates this picture, which corresponds to applying the fundamental theorem of calculus, one obtains Cavalieri's quadrature formula, the integral $\int x^{n-1}\,dx = \tfrac{1}{n}x^n$ – see proof of Cavalieri's quadrature formula for details.

Binomial Coefficients

The coefficients that appear in the binomial expansion are called binomial coefficients. These are usually written $\tbinom{n}{k}$, and pronounced "n choose k".

Formulae

The coefficient of $x^{n-k}y^k$ is given by the formula;

$$\binom{n}{k} = \frac{n!}{k!(n-k)!}$$

which is defined in terms of the factorial function $n!$. Equivalently, this formula can be written

$$\binom{n}{k} = \frac{n(n-1)\cdots(n-k+1)}{k(k-1)\cdots 1} = \prod_{\ell=1}^{k}\frac{n-\ell+1}{\ell} = \prod_{\ell=0}^{k-1}\frac{n-\ell}{k-\ell}$$

with k factors in both the numerator and denominator of the fraction. Note that, although this formula involves a fraction, the binomial coefficient $\tbinom{n}{k}$ is actually an integer.

Combinatorial Interpretation

The binomial coefficient $\tbinom{n}{k}$ can be interpreted as the number of ways to choose k elements from an n-element set. This is related to binomials for the following reason: if we write $(x + y)^n$ as a product

$$(x+y)(x+y)(x+y)\cdots(x+y),$$

then, according to the distributive law, there will be one term in the expansion for each choice of either x or y from each of the binomials of the product. For example, there will only be one term x^n, corresponding to choosing x from each binomial. However, there will be several terms of the form $x^{n-2}y^2$, one for each way of choosing exactly two binomials to contribute a y. Therefore, after combining like terms, the coefficient of $x^{n-2}y^2$ will be equal to the number of ways to choose exactly 2 elements from an n-element set.

Proofs

Combinatorial Proof

Example

The coefficient of xy^2 in:

$$\begin{aligned}(x+y)^3 &= (x+y)(x+y)(x+y)\\ &= xxx + xxy + xyx + \underline{xyy} + yxx + \underline{yxy} + \underline{yyx} + yyy\\ &= x^3 + 3x^2y + \underline{3xy^2} + y^3.\end{aligned}$$

equals $\binom{3}{2} = 3$ because there are three x,y strings of length 3 with exactly two y's, namely,

$$xyy,\ yxy,\ yyx,$$

corresponding to the three 2-element subsets of { 1, 2, 3 }, namely,

$$\{2,3\},\ \{1,3\},\ \{1,2\},$$

where each subset specifies the positions of the y in a corresponding string.

General Case

Expanding $(x + y)^n$ yields the sum of the 2^n products of the form $e_1e_2 \dots e_n$ where each e_i is x or y. Rearranging factors shows that each product equals $x^{n-k}y^k$ for some k between 0 and n. For a given k, the following are proved equal in succession:

- the number of copies of $x^{n-k}y^k$ in the expansion
- the number of n-character x,y strings having y in exactly k positions
- the number of k-element subsets of $\{1, 2, \dots, n\}$
- $\binom{n}{k}$ (this is either by definition, or by a short combinatorial argument if one is defining $\binom{n}{k}$ as $\frac{n!}{k!(n-k)!}$).

This proves the binomial theorem.

Inductive Proof

Induction yields another proof of the binomial theorem. When $n = 0$, both sides equal 1, since $x^0 = 1$ and $\binom{0}{0} = 1$. Now suppose that the equality holds for a given n; we will prove it for $n + 1$. For $j, k \geq 0$, let $[f(x, y)]_{j,k}$ denote the coefficient of $x^j y^k$ in the polynomial $f(x, y)$. By the inductive hypothesis, $(x + y)^n$ is a polynomial in x and y such that $[(x + y)^n]_{j,k}$ is $\binom{n}{k}$ if $j + k = n$, and 0 otherwise. The identity;

$$(x+y)^{n+1} = x(x+y)^n + y(x+y)^n$$

shows that $(x + y)^{n+1}$ also is a polynomial in x and y, and

$$[(x+y)^{n+1}]_{j,k} = [(x+y)^n]_{j-1,k} + [(x+y)^n]_{j,k-1},$$

since if $j + k = n + 1$, then $(j - 1) + k = n$ and $j + (k - 1) = n$. Now, the right hand side is

$$\binom{n}{k} + \binom{n}{k-1} = \binom{n+1}{k},$$

by Pascal's identity. On the other hand, if $j + k \neq n + 1$, then $(j - 1) + k \neq n$ and $j + (k - 1) \neq n$, so we get $0 + 0 = 0$. Thus;

$$(x+y)^{n+1} = \sum_{k=0}^{n+1} \binom{n+1}{k} x^{n+1-k} y^k,$$

which is the inductive hypothesis with $n + 1$ substituted for n and so completes the inductive step.

Generalizations

Newton's Generalized Binomial Theorem

Around 1665, Isaac Newton generalized the binomial theorem to allow real exponents other than nonnegative integers. (The same generalization also applies to complex exponents.) In this generalization, the finite sum is replaced by an infinite series. In order to do this, one needs to give meaning to binomial coefficients with an arbitrary upper index, which cannot be done using the usual formula with factorials. However, for an arbitrary number r, one can define;

$$\binom{r}{k} = \frac{r(r-1)\cdots(r-k+1)}{k!} = \frac{(r)_k}{k!},$$

where $(\cdot)_k$ is the Pochhammer symbol, here standing for a falling factorial. This agrees with the usual definitions when r is a nonnegative integer. Then, if x and y are real numbers with $|x| > |y|$,[Note 1] and r is any complex number, one has;

$$\begin{aligned}(x+y)^r &= \sum_{k=0}^{\infty} \binom{r}{k} x^{r-k} y^k \\ &= x^r + rx^{r-1}y + \frac{r(r-1)}{2!}x^{r-2}y^2 + \frac{r(r-1)(r-2)}{3!}x^{r-3}y^3 + \cdots.\end{aligned}$$

When r is a nonnegative integer, the binomial coefficients for $k > r$ are zero, so this equation reduces to the usual binomial theorem, and there are at most $r + 1$ nonzero terms. For other values of r, the series typically has infinitely many nonzero terms.

For example, with $r = 1/2$ gives the following series for the square root:

$$\sqrt{1+x} = 1 + \frac{1}{2}x - \frac{1}{8}x^2 + \frac{1}{16}x^3 - \frac{5}{128}x^4 + \frac{7}{256}x^5 - \cdots$$

Taking $r = -1$, the generalized binomial series gives the geometric series formula, valid for $|x| < 1$:

$$(1+x)^{-1} = \frac{1}{1+x} = 1 - x + x^2 - x^3 + x^4 - x^5 + \cdots$$

More generally, with $r = -s$:

$$\frac{1}{(1-x)^s} = \sum_{k=0}^{\infty} \binom{s+k-1}{k} x^k \equiv \sum_{k=0}^{\infty} \binom{s+k-1}{s-1} x^k.$$

So, for instance, when $s = 1/2$,

$$\frac{1}{\sqrt{1+x}} = 1 - \frac{1}{2}x + \frac{3}{8}x^2 - \frac{5}{16}x^3 + \frac{35}{128}x^4 - \frac{63}{256}x^5 + \cdots$$

Further Generalizations

The generalized binomial theorem can be extended to the case where x and y are complex numbers. For this version, one should again assume $|x| > |y|$ and define the powers of $x + y$ and x using a holomorphic branch of log defined on an open disk of radius $|x|$ centered at x.

The generalized binomial theorem is valid also for elements x and y of a Banach algebra as long as $xy = yx$, x is invertible, and $||y/x|| < 1$.

Multinomial Theorem

The binomial theorem can be generalized to include powers of sums with more than two terms. The general version is;

$$(x_1 + x_2 + \cdots + x_m)^n = \sum_{k_1+k_2+\cdots+k_m=n} \binom{n}{k_1, k_2, \ldots, k_m} x_1^{k_1} x_2^{k_2} \cdots x_m^{k_m}.$$

where the summation is taken over all sequences of nonnegative integer indices k_1 through k_m such that the sum of all k_i is n. (For each term in the expansion, the exponents must add up to n). The coefficients $\binom{n}{k_1,\cdots,k_m}$ are known as multinomial coefficients, and can be computed by the formula;

$$\binom{n}{k_1, k_2, \ldots, k_m} = \frac{n!}{k_1! \cdot k_2! \cdots k_m!}.$$

Combinatorially, the multinomial coefficient $\binom{n}{k_1,\cdots,k_m}$ counts the number of different ways to partition an n-element set into disjoint subsets of sizes $k_1, ..., k_m$.

Multi-binomial Theorem

It is often useful when working in more dimensions, to deal with products of binomial expressions. By the binomial theorem this is equal to;

$$(x_1+y_1)^{n_1}\cdots(x_d+y_d)^{n_d}=\sum_{k_1=0}^{n_1}\cdots\sum_{k_d=0}^{n_d}\binom{n_1}{k_1}x_1^{k_1}y_1^{n_1-k_1}\ldots\binom{n_d}{k_d}x_d^{k_d}y_d^{n_d-k_d}.$$

This may be written more concisely, by multi-index notation, as;

$$(x+y)^{\alpha}=\sum_{\nu\le\alpha}\binom{\alpha}{\nu}x^{\nu}y^{\alpha-\nu}.$$

Applications

Multiple-angle Identities

For the complex numbers the binomial theorem can be combined with De Moivre's formula to yield multiple-angle formulas for the sine and cosine. According to De Moivre's formula,

$$\cos(nx)+i\sin(nx)=(\cos x+i\sin x)^n.$$

Using the binomial theorem, the expression on the right can be expanded, and then the real and imaginary parts can be taken to yield formulas for cos(nx) and sin(nx). For example, since,

$$(\cos x+i\sin x)^2=\cos^2 x+2i\cos x\sin x-\sin^2 x,$$

De Moivre's formula tells us that,

$$\cos(2x)=\cos^2 x-\sin^2 x \quad \text{and} \quad \sin(2x)=2\cos x\sin x,$$

which are the usual double-angle identities. Similarly, since,

$$(\cos x+i\sin x)^3=\cos^3 x+3i\cos^2 x\sin x-3\cos x\sin^2 x-i\sin^3 x,$$

De Moivre's formula yields,

$$\cos(3x)=\cos^3 x-3\cos x\sin^2 x \quad \text{and} \quad \sin(3x)=3\cos^2 x\sin x-\sin^3 x.$$

In general,

$$\cos(nx) = \sum_{k \text{ even}} (-1)^{k/2} \binom{n}{k} \cos^{n-k} x \sin^k x$$

and

$$\sin(nx) = \sum_{k \text{ odd}} (-1)^{(k-1)/2} \binom{n}{k} \cos^{n-k} x \sin^k x.$$

Series for e

The number e is often defined by the formula:

$$e = \lim_{n\to\infty} \left(1 + \frac{1}{n}\right)^n.$$

Applying the binomial theorem to this expression yields the usual infinite series for e. In particular:

$$\left(1+\frac{1}{n}\right)^n = 1 + \binom{n}{1}\frac{1}{n} + \binom{n}{2}\frac{1}{n^2} + \binom{n}{3}\frac{1}{n^3} + \cdots + \binom{n}{n}\frac{1}{n^n}.$$

The kth term of this sum is:

$$\binom{n}{k}\frac{1}{n^k} = \frac{1}{k!}\cdot\frac{n(n-1)(n-2)\cdots(n-k+1)}{n^k}$$

As $n \to \infty$, the rational expression on the right approaches one, and therefore:

$$\lim_{n\to\infty}\binom{n}{k}\frac{1}{n^k} = \frac{1}{k!}.$$

This indicates that e can be written as a series:

$$e = \sum_{k=0}^{\infty}\frac{1}{k!} = \frac{1}{0!} + \frac{1}{1!} + \frac{1}{2!} + \frac{1}{3!} + \cdots.$$

Indeed, since each term of the binomial expansion is an increasing function of n, it follows from the monotone convergence theorem for series that the sum of this infinite series is equal to e.

Derivative of the Power Function

In finding the derivative of the power function $f(x) = x^n$ for integer n using the definition of derivative, one can expand the binomial $(x + h)^n$.

Nth Derivative of a Product

To indicate the formula for the derivative of order n of the product of two functions, the formula of the binomial theorem is used symbolically.

Probability

The binomial theorem is closely related to the probability mass function of the Negative binomial distribution. The probability of a (countable) collection of independent Bernoulli trials $\{X_t\}_{t\in S}$ with probability of success $p \in [0,1]$ all not happening is:

$$P(\cap_{t\in S} X_t^C) = (1-p)^{|S|} = \sum_{n=0}^{|S|} \binom{|S|}{n} (-p)^n$$

A useful upper bound for this quantity is e^{-pn}.

The Binomial Theorem in Abstract Algebra

Formula (1) is valid more generally for any elements x and y of a semiring satisfying $xy = yx$. The theorem is true even more generally: alternativity suffices in place of associativity.

The binomial theorem can be stated by saying that the polynomial sequence { 1, x, x^2, x^3, ... } is of binomial type.

In Popular Culture

- The binomial theorem is mentioned in the Major-General's Song in the comic opera The Pirates of Penzance.
- Professor Moriarty is described by Sherlock Holmes as having written a treatise on the binomial theorem.
- The Portuguese poet Fernando Pessoa, using the heteronym Álvaro de Campos, wrote that "Newton's Binomial is as beautiful as the Venus de Milo. The truth is that few people notice it."

Chevalley–Warning Theorem

In algebra, the Chevalley–Warning theorem implies that certain polynomial equations in sufficiently many variables over a finite field have solutions. It was proved by Ewald Warning (1935) and a slightly weaker form of the theorem, known as Chevalley's theorem, was proved by Chevalley (1935). Chevalley's theorem implied Artin's and Dickson's conjecture that finite fields are quasi-algebraically closed fields (Artin 1982, page x).

Statement of the Theorems

Let $\mathbb{F}$ be a finite field and $\{f_j\}_{j=1}^r \subseteq \mathbb{F}[X_1,\ldots,X_n]$ be a set of polynomials such that the number of variables satisfies:

$$n > \sum_{j=1}^{r} d_j$$

where d_j is the total degree of f_j. The theorems are statements about the solutions of the following system of polynomial equations:

$$f_j(x_1,\ldots,x_n) = 0 \quad \text{for } j = 1,\ldots,r.$$

- *Chevalley–Warning theorem* states that the number of common solutions is divisible by the characteristic p of $\mathbb{F}$. Or in other words, the cardinality of the vanishing set of $\{f_j\}_{j=1}^{r}$ is 0 modulo p.
- *Chevalley's theorem* states that if the system has the trivial solution $(0,\ldots,0) \in \mathbb{F}^n$, i.e. if the polynomials have no constant terms, then the system also has a non-trivial solution $(a_1,\ldots,a_n) \in \mathbb{F}^n \setminus \{(0,\ldots,0)\}$.

Chevalley's theorem is an immediate consequence of the Chevalley–Warning theorem since p is at least 2.

Both theorems are best possible in the sense that, given any n, the list $f_j = x_j, j = 1,\ldots,n$ has total degree n and only the trivial solution. Alternatively, using just one polynomial, we can take f_1 to be the degree n polynomial given by the norm of $x_1a_1 + \ldots + x_na_n$ where the elements a form a basis of the finite field of order p^n.

Warning proved another theorem, known as Warning's second theorem, which states that if the system of polynomial equations has the trivial solution, then it has at least q^{n-d} solutions where q is the size of the finite field and $d := d_1 + \ldots + d_r$. Chevalley's theorem also follows directly from this.

Proof of Warning's Theorem

Remark: If $i < q-1$ then

$$\sum_{x \in \mathbb{F}} x^i = 0$$

so the sum over $\mathbb{F}^n$ of any polynomial in $x_1,\ldots,x_n$ of degree less than $n(q-1)$ also vanishes.

The total number of common solutions modulo p of $f_1,\ldots,f_r = 0$ is equal to:

$$\sum_{x \in \mathbb{F}^n} (1 - f_1^{q-1}(x)) \cdot \ldots \cdot (1 - f_r^{q-1}(x))$$

because each term is 1 for a solution and 0 otherwise. If the sum of the degrees of the polynomials f_i is less than n then this vanishes by the remark above.

Artin's Conjecture

It is a consequence of Chevalley's theorem that finite fields are quasi-algebraically closed. This had been conjectured by Emil Artin in 1935. The motivation behind Artin's conjecture was his observation that quasi-algebraically closed fields have trivial Brauer group, together with the fact that finite fields have trivial Brauer group by Wedderburn's theorem.

The Ax–Katz Theorem

The Ax–Katz theorem, named after James Ax and Nicholas Katz, determines more accurately a power q^b of the cardinality q of $\mathbb{F}$ dividing the number of solutions; here, if d is the largest of the d_j, then the exponent b can be taken as the ceiling function of:

$$\frac{n-\sum_j d_j}{d}.$$

The Ax–Katz result has an interpretation in étale cohomology as a divisibility result for the (reciprocals of) the zeroes and poles of the local zeta-function. Namely, the same power of divides each of these algebraic integers.

Boolean Prime Ideal Theorem

In mathematics, a prime ideal theorem guarantees the existence of certain types of subsets in a given algebra. A common example is the Boolean prime ideal theorem, which states that ideals in a Boolean algebra can be extended to prime ideals. A variation of this statement for filters on sets is known as the ultrafilter lemma. Other theorems are obtained by considering different mathematical structures with appropriate notions of ideals, for example, rings and prime ideals (of ring theory), or distributive lattices and *maximal* ideals (of order theory). This article focuses on prime ideal theorems from order theory.

Although the various prime ideal theorems may appear simple and intuitive, they cannot be derived in general from the axioms of Zermelo–Fraenkel set theory without the axiom of choice (abbreviated ZF). Instead, some of the statements turn out to be equivalent to the axiom of choice (AC), while others—the Boolean prime ideal theorem, for instance—represent a property that is strictly weaker than AC. It is due to this intermediate status between ZF and ZF + AC (ZFC) that the Boolean prime ideal theorem is often taken as an axiom of set theory. The abbreviations BPI or PIT (for Boolean algebras) are sometimes used to refer to this additional axiom.

Prime Ideal Theorems

Recall that an order ideal is a (non-empty) directed lower set. If the considered poset has binary suprema (a.k.a. joins), as do the posets within this article, then this is equivalently characterized as a non-empty lower set I that is closed for binary suprema (i.e. x, y in I imply xy in I). An ideal I is prime if its set-theoretic complement in the poset is a filter. Ideals are proper if they are not equal to the whole poset.

Historically, the first statement relating to later prime ideal theorems was in fact referring to filters—subsets that are ideals with respect to the dual order. The ultrafilter lemma states that every filter on a set is contained within some maximal (proper) filter—an *ultrafilter*. Recall that filters on sets are proper filters of the Boolean algebra of its powerset. In this special case, maximal filters (i.e. filters that are not strict subsets of any proper filter) and prime filters (i.e. filters that with each union of subsets X and Y contain also X or Y) coincide. The dual of this statement thus assures that every ideal of a powerset is contained in a prime ideal.

The above statement led to various generalized prime ideal theorems, each of which exists in a weak and in a strong form. *Weak prime ideal theorems* state that every *non-trivial* algebra of a certain class has at least one prime ideal. In contrast, *strong prime ideal theorems* require that every ideal that is disjoint from a given filter can be extended to a prime ideal that is still disjoint from that filter. In the case of algebras that are not posets, one uses different substructures instead of filters. Many forms of these theorems are actually known to be equivalent, so that the assertion that "PIT" holds is usually taken as the assertion that the corresponding statement for Boolean algebras (BPI) is valid.

Another variation of similar theorems is obtained by replacing each occurrence of *prime ideal* by *maximal ideal*. The corresponding maximal ideal theorems (MIT) are often—though not always—stronger than their PIT equivalents.

Boolean Prime Ideal Theorem

The Boolean prime ideal theorem is the strong prime ideal theorem for Boolean algebras. Thus the formal statement is:

> Let B be a Boolean algebra, let I be an ideal and let F be a filter of B, such that I and F are disjoint. Then I is contained in some prime ideal of B that is disjoint from F.

The weak prime ideal theorem for Boolean algebras simply states:

> Every Boolean algebra contains a prime ideal.

We refer to these statements as the weak and strong *BPI*. The two are equivalent, as the strong BPI clearly implies the weak BPI, and the reverse implication can be achieved by using the weak BPI to find prime ideals in the appropriate quotient algebra.

The BPI can be expressed in various ways. For this purpose, recall the following theorem:

For any ideal I of a Boolean algebra B, the following are equivalent:

- I is a prime ideal.
- I is a maximal ideal, i.e. for any proper ideal J, if I is contained in J then $I = J$.
- For every element a of B, I contains exactly one of $\{a, \neg a\}$.

This theorem is a well-known fact for Boolean algebras. Its dual establishes the equivalence of prime filters and ultrafilters. Note that the last property is in fact self-dual—only the prior assumption that I is an ideal gives the full characterization. All of the implications within this theorem can be proven in ZF.

Thus the following (strong) maximal ideal theorem (MIT) for Boolean algebras is equivalent to BPI:

> Let B be a Boolean algebra, let I be an ideal and let F be a filter of B, such that I and F are disjoint. Then I is contained in some maximal ideal of B that is disjoint from F.

Note that one requires "global" maximality, not just maximality with respect to being disjoint from F. Yet, this variation yields another equivalent characterization of BPI:

> Let B be a Boolean algebra, let I be an ideal and let F be a filter of B, such that I and F are disjoint. Then I is contained in some ideal of B that is maximal among all ideals disjoint from F.

The fact that this statement is equivalent to BPI is easily established by noting the following theorem: For any distributive lattice L, if an ideal I is maximal among all ideals of L that are disjoint to a given filter F, then I is a prime ideal. The proof for this statement (which can again be carried out in ZF set theory) is included in the article on ideals. Since any Boolean algebra is a distributive lattice, this shows the desired implication.

All of the above statements are now easily seen to be equivalent. Going even further, one can exploit the fact the dual orders of Boolean algebras are exactly the Boolean algebras themselves. Hence, when taking the equivalent duals of all former statements, one ends up with a number of theorems that equally apply to Boolean algebras, but where every occurrence of *ideal* is replaced by *filter*. It is worth noting that for the special case where the Boolean algebra under consideration is a powerset with the subset ordering, the "maximal filter theorem" is called the ultrafilter lemma.

Summing up, for Boolean algebras, the weak and strong MIT, the weak and strong PIT, and these statements with filters in place of ideals are all equivalent. It is known that all of these statements are consequences of the Axiom of Choice, *AC*, (the easy proof makes use of Zorn's lemma), but cannot be proven in ZF (Zermelo-Fraenkel set theory without *AC*), if ZF is consistent. Yet, the BPI is strictly weaker than the axiom of choice, though the proof of this statement, due to J. D. Halpern and Azriel Lévy is rather non-trivial.

Further Prime Ideal Theorems

The prototypical properties that were discussed for Boolean algebras in the above section can easily be modified to include more general lattices, such as distributive lattices or Heyting algebras. However, in these cases maximal ideals are different from prime ideals, and the relation between PITs and MITs is not obvious.

Indeed, it turns out that the MITs for distributive lattices and even for Heyting algebras are equivalent to the axiom of choice. On the other hand, it is known that the strong PIT for distributive lattices is equivalent to BPI (i.e. to the MIT and PIT for Boolean algebras). Hence this statement is strictly weaker than the axiom of choice. Furthermore, observe that Heyting algebras are not self dual, and thus using filters in place of ideals yields different theorems in this setting. Maybe surprisingly, the MIT for the duals of Heyting algebras is not stronger than BPI, which is in sharp contrast to the abovementioned MIT for Heyting algebras.

Finally, prime ideal theorems do also exist for other (not order-theoretical) abstract algebras. For example, the MIT for rings implies the axiom of choice. This situation requires to replace the order-theoretic term "filter" by other concepts—for rings a "multiplicatively closed subset" is appropriate.

The Ultrafilter Lemma

A filter on a set X is a nonempty collection of nonempty subsets of X that is closed under finite intersection and under superset. An ultrafilter is a maximal filter. The ultrafilter lemma states that every filter on a set X is a subset of some ultrafilter on X. This lemma is most often used in the study of topology. An ultrafilter that does not contain finite sets is called non-principal. The ultrafilter lemma, and in particular the existence of non-principal ultrafilters (consider the filter of all sets with finite complements), follows easily from Zorn's lemma.

The ultrafilter lemma is equivalent to the Boolean prime ideal theorem, with the equivalence provable in ZF set theory without the axiom of choice. The idea behind the proof is that the subsets of any set form a Boolean algebra partially ordered by inclusion, and any Boolean algebra is representable as an algebra of sets by Stone's representation theorem.

Applications

Intuitively, the Boolean prime ideal theorem states that there are "enough" prime ideals in a Boolean algebra in the sense that we can extend *every* ideal to a maximal one. This is of practical importance for proving Stone's representation theorem for Boolean algebras, a special case of Stone duality, in which one equips the set of all prime ideals with a certain topology and can indeed regain the original Boolean algebra (up to isomorphism) from this data. Furthermore, it turns out that in applications one can freely choose either to work with prime ideals or with prime filters, because every ideal uniquely determines a filter: the set of all Boolean complements of its elements. Both approaches are found in the literature.

Many other theorems of general topology that are often said to rely on the axiom of choice are in fact equivalent to BPI. For example, the theorem that a product of compact Hausdorff spaces is compact is equivalent to it. If we leave out "Hausdorff" we get a theorem equivalent to the full axiom of choice.

A not too well known application of the Boolean prime ideal theorem is the existence of a non-measurable set (the example usually given is the Vitali set, which requires the axiom of choice). From this and the fact that the BPI is strictly weaker than the axiom of choice, it follows that the existence of non-measurable sets is strictly weaker than the axiom of choice.

In linear algebra, the Boolean prime ideal theorem can be used to prove that any two bases of a given vector space have the same cardinality.

Rational Root Theorem

In algebra, the rational root theorem (or rational root test, rational zero theorem, rational zero test or p/q theorem) states a constraint on rational solutions of a polynomial equation.

$$a_n x^n + a_{n-1} x^{n-1} + \cdots + a_0 = 0$$

with integer coefficients. These solutions are the possible roots (equivalently, zeroes) of the polynomial on the left side of the equation.

If a_0 and a_n are nonzero, then each rational solution x, when written as a fraction $x = p/q$ in lowest terms (i.e., the greatest common divisor of p and q is 1), satisfies:

- p is an integer factor of the constant term a_0, and
- q is an integer factor of the leading coefficient a_n.

The rational root theorem is a special case (for a single linear factor) of Gauss's lemma on the factorization of polynomials. The integral root theorem is a special case of the rational root theorem if the leading coefficient $a_n = 1$.

Application

The theorem is used in order to determine whether a polynomial has any rational roots, and if so to find them. Since the theorem gives constraints on the numerator and denominator of the fully reduced rational roots as being divisors of certain numbers, all possible combinations of divisors can be checked and either the rational roots will be found or it will be determined that there are none. If one or more are found, they can be factored out of the polynomial, resulting in a polynomial of lower degree whose roots are also roots of the original polynomial.

Cubic Equation

The general cubic equation;

$$ax^3 + bx^2 + cx + d = 0$$

with integer coefficients has three solutions in the complex plane. If it is found by the rational root test that there are no rational solutions, then the only way to express the solutions algebraically is to use cube roots. But if the test finds three rational solutions, then the cube roots are avoided. And if precisely one rational solution r is found to exist, then $(x - r)$ can be factored out of the cubic polynomial using polynomial long division, leaving a quadratic polynomial whose two roots are the remaining two roots of the cubic; and these can be found using the quadratic formula, again avoiding the use of cube roots.

Proofs

First Proof

Let $P(x) = a_nx^n + a_{n-1}x^{n-1} + \ldots + a_1x + a_0$ for some $a_0, \ldots, a_n \in Z$, and suppose $P(p/q) = 0$ for some coprime $p, q \in Z$:

$$P\left(\tfrac{p}{q}\right) = a_n\left(\tfrac{p}{q}\right)^n + a_{n-1}\left(\tfrac{p}{q}\right)^{n-1} + \cdots + a_1\left(\tfrac{p}{q}\right) + a_0 = 0.$$

If we multiply both sides by q^n, shift the constant term to the right hand side, and factor out p on the left hand side, we get;

$$p(a_np^{n-1} + a_{n-1}qp^{n-2} + \cdots + a_1q^{n-1}) = -a_0q^n.$$

We see that p times the integer quantity in parentheses equals $-a_0q^n$, so p divides a_0q^n. But p is coprime to q and therefore to q^n, so by (the generalized form of) Euclid's lemma it must divide the remaining factor a_0 of the product.

If we instead shift the leading term to the right hand side and factor out q on the left hand side, we get;

$$q(a_{n-1}p^{n-1} + a_{n-2}qp^{n-2} + \cdots + a_0 q^{n-1}) = -a_n p^n.$$

And for similar reasons, we can conclude that q divides a_n.

Proof using Gauss's Lemma

Should there be a nontrivial factor dividing all the coefficients of the polynomial, then one can divide by the greatest common divisor of the coefficients so as to obtain a primitive polynomial in the sense of Gauss's lemma; this does not alter the set of rational roots and only strengthens the divisibility conditions. That lemma says that if the polynomial factors in Q[X], then it also factors in Z[X] as a product of primitive polynomials. Now any rational root p/q corresponds to a factor of degree 1 in Q[X] of the polynomial, and its primitive representative is then $qx - p$, assuming that p and q are coprime. But any multiple in Z[X] of $qx - p$ has leading term divisible by q and constant term divisible by p, which proves the statement. This argument shows that more generally, any irreducible factor of P can be supposed to have integer coefficients, and leading and constant coefficients dividing the corresponding coefficients of P.

Examples

First

In the polynomial

$$2x^3 + x - 1,$$

any rational root fully reduced would have to have a numerator that divides evenly into 1 and a denominator that divides evenly into 2. Hence the only possible rational roots are ±1/2 and ±1; since neither of these equates the polynomial to zero, it has no rational roots.

Second

In the polynomial

$$x^3 - 7x + 6$$

the only possible rational roots would have a numerator that divides 6 and a denominator that divides 1, limiting the possibilities to ±1, ±2, ±3, and ±6. of these, 1, 2, and −3 equate the polynomial to zero, and hence are its rational roots. (In fact these are its only roots since a cubic has only three roots; in general a polynomial could have some rational and some irrational roots.)

Third

Every rational root of the polynomial

$$3x^3 - 5x^2 + 5x - 2$$

must be among the numbers symbolically indicated by:

$$\pm\frac{1,2}{1,3},$$

which gives the list of 8 possible answers:

$$\pm\{1,2,\frac{1}{3},\frac{2}{3}\}.$$

These root candidates can be tested using Horner's method (for instance). In this particular case there is exactly one rational root. If a root candidate does not cause the polynomial to equal zero, it can be used to shorten the list of remaining candidates. For example, $x = 1$ does not work, as the polynomial then equals 1. This means that substituting $x = 1 + t$ yields a polynomial in t with constant term 1, while the coefficient of t^3 remains the same as the coefficient of x^3. Applying the rational root theorem thus yields the following possible roots for t:

$$t=\pm\frac{1}{1,3}.$$

Therefore,

$$x=1+t=2,0,\frac{4}{3},\frac{2}{3}.$$

Root candidates that do not occur on both lists are ruled out. The list of rational root candidates has thus shrunk to just $x = 2$ and $x = 2/3$.

If k rational roots are found ($k \geq 1$), Horner's method will also yield a polynomial of degree $n - k$ whose roots, together with the rational roots, are exactly the roots of the original polynomial. It may also be the case that none of the candidates is a solution; in this case the equation setting the polynomial equal to 0 has no rational solution. If the equation lacks a constant term a_0, then 0 is one of the rational solutions of the equation.

Factor Theorem

In algebra, the factor theorem is a theorem linking factors and zeros of a polynomial. It is a special case of the polynomial remainder theorem.

The factor theorem states that a polynomial $f(x)$ has a factor $(x-k)$ if and only if $f(x) = 0$(i.e. k is a root).

Factorization of Polynomials

Two problems where the factor theorem is commonly applied are those of factoring a polynomial and finding the roots of a polynomial equation; it is a direct consequence of the theorem that these problems are essentially equivalent.

The factor theorem is also used to remove known zeros from a polynomial while leaving all unknown zeros intact, thus producing a lower degree polynomial whose zeros may be easier to find. Abstractly, the method is as follows:

1. "Guess" a zero a of the polynomial f. (In general, this can be *very hard*, but maths textbook problems that involve solving a polynomial equation are often designed so that some roots are easy to discover.)
2. Use the factor theorem to conclude that $(x-a)$ is a factor of $f(x)$.
3. Compute the polynomial $g(x) = f(x)/(x-a)$, for example using polynomial long division or synthetic division.
4. Conclude that any root $x \neq a$ of $f(x) = 0$ is a root of $g(x) = 0$. Since the polynomial degree of g is one less than that of f, it is "simpler" to find the remaining zeros by studying g.

Example

Find the factors at

$$x^3 + 7x^2 + 8x + 2.$$

To do this you would use trial and error (or the rational root theorem) to find the first x value that causes the expression to equal zero. To find out if $(x-1)$ is a factor, substitute $x = 1$ into the polynomial above:

$$x^3 + 7x^2 + 8x + 2 = (1)^3 + 7(1)^2 + 8(1) + 2$$

$$= 1 + 7 + 8 + 2$$

$$= 18.$$

As this is equal to 18 and not 0 this means $(x-1)$ is not a factor of $x^3 + 7x^2 + 8x + 2$. So, we next try $(x+1)$ (substituting $x = -1$ into the polynomial):

$$(-1)^3 + 7(-1)^2 + 8(-1) + 2.$$

This is equal to 0. Therefore $x-(-1)$, which is to say $x+1$, is a factor, and -1 is a root of $x^3 + 7x^2 + 8x + 2$.

The next two roots can be found by algebraically dividing $x^3 + 7x^2 + 8x + 2$ by $(x+1)$ to get a quadratic, which can be solved directly, by the factor theorem or by the quadratic formula.

$$\frac{x^3 + 7x^2 + 8x + 2}{x+1} = x^2 + 6x + 2$$

and therefore $(x+1)$ and $x^2 + 6x + 2$ are the factors of $x^3 + 7x^2 + 8x + 2$.

In algebra, the Krull–Akizuki theorem states the following: let *A* be a one-dimensional reduced noetherian ring, *K* its total ring of fractions. If *B* is a subring of a finite extension *L* of *K* containing *A* and is not a field, then *B* is a one-dimensional noetherian ring. Furthermore, for every nonzero ideal *I* of *B*, $B >$is finite over *A*.

Note that the theorem does not say that *B* is finite over *A*. The theorem does not extend to higher dimension. One important consequence of the theorem is that the integral closure of a Dedekind domain *A* in a finite extension of the field of fractions of *A* is again a Dedekind domain. This consequence does generalize to a higher dimension: the Mori–Nagata theorem states that the integral closure of a noetherian domain is a Krull domain.

Proof

Here, we give a proof when $L = K$. Let $\mathfrak{p}_i$ be minimal prime ideals of *A*; there are finitely many of them. Let K_i be the field of fractions of $A/\mathfrak{p}_i$ and I_i the kernel of the natural map $B \to K \to K_i$. Then we have:

$$A/\mathfrak{p}_i \subset B/I_i \subset K_i.$$

Now, if the theorem holds when *A* is a domain, then this implies that *B* is a one-dimensional noetherian domain since each B/I_i is and since $B = \prod B/I_i$. Hence, we reduced the proof to the case *A* is a domain. Let $0 \neq I \subset B$ be an ideal and let a be a nonzero element in the nonzero ideal $I \cap A$. Set $I_n = a^n B \cap A + aA$. Since $A >$is a zero-dim noetherian ring; thus, artinian, there is an l such that $I_n = I_l$ for all $n \geq l$. We claim

$$a^l B \subset a^{l+1} B + A.$$

Since it suffices to establish the inclusion locally, we may assume *A* is a local ring with the maximal ideal $\mathfrak{m}$. Let *x* be a nonzero element in *B*. Then, since *A* is noetherian, there is an *n* such that $\mathfrak{m}^{n+1} \subset x^{-1}A$ and so $a^{n+1}x \in a^{n+1}B \cap A \subset I_{n+2}$. Thus,

$$a^n x \in a^{n+1} B \cap A + A.$$

Now, assume *n* is a minimum integer such that $n \geq l$ and the last inclusion holds. If , then we easily see that $a^n x \in I_{n+1}$. But then the above inclusion holds for $n-1$, contradiction. Hence, we have $n = l$ and this establishes the claim. It now follows:

$$B/aB \simeq a^l B / a^{l+1} B \subset (a^{l+1}B + A)/a^{l+1}B \simeq A/a^{l+1}B \cap A.$$

Hence, $B >$has finite length as *A*-module. In particular, the image of *I* there is finitely generated and so *I* is finitely generated. Finally, the above shows that $B >$has zero dimension and so *B* has dimension one.

References

- Bourbaki, N. (18 November 1998). Elements of the History of Mathematics Paperback. J. Meldrum (Translator). ISBN 978-3-540-64767-6

- Alekseev, V. B. (2004), Abel's theorem in problems and solutions. Based on the lectures of Professor V. I. Arnold, Kluwer Academic Publishers, ISBN 1-4020-2186-0
- Pesic, Peter (2004), Abel's proof. An essay on the sources and meaning of mathematical unsolvability, MIT Press, ISBN 0-262-66182-9
- Tignol, Jean-Pierre (2016), "Galois", Galois' Theory of algebraic equations (2nd ed.), World Scientific, ISBN 978-981-4704-69-4
- Cover, Thomas M.; Thomas, Joy A. (2001-01-01). Data Compression. John Wiley & Sons, Inc. p. 320. doi:10.1002/0471200611.ch5. ISBN 9780471200611
- Khovanskii, Askold (2014), Topological Galois Theory: Solvability and Unsolvability of Equations in Finite Terms, Springer Monographs in Mathematics, Springer-Verlag, doi:10.1007/978-3-642-38871-2, ISBN 978-3-642-38870-5
- Tignol, Jean-Pierre (2016), "Ruffini and Abel on general equations", Galois' Theory of algebraic equations (2nd ed.), World Scientific, ISBN 978-981-4704-69-4

Permissions

All chapters in this book are published with permission under the Creative Commons Attribution Share Alike License or equivalent. Every chapter published in this book has been scrutinized by our experts. Their significance has been extensively debated. The topics covered herein carry significant information for a comprehensive understanding. They may even be implemented as practical applications or may be referred to as a beginning point for further studies.

We would like to thank the editorial team for lending their expertise to make the book truly unique. They have played a crucial role in the development of this book. Without their invaluable contributions this book wouldn't have been possible. They have made vital efforts to compile up to date information on the varied aspects of this subject to make this book a valuable addition to the collection of many professionals and students.

This book was conceptualized with the vision of imparting up-to-date and integrated information in this field. To ensure the same, a matchless editorial board was set up. Every individual on the board went through rigorous rounds of assessment to prove their worth. After which they invested a large part of their time researching and compiling the most relevant data for our readers.

The editorial board has been involved in producing this book since its inception. They have spent rigorous hours researching and exploring the diverse topics which have resulted in the successful publishing of this book. They have passed on their knowledge of decades through this book. To expedite this challenging task, the publisher supported the team at every step. A small team of assistant editors was also appointed to further simplify the editing procedure and attain best results for the readers.

Apart from the editorial board, the designing team has also invested a significant amount of their time in understanding the subject and creating the most relevant covers. They scrutinized every image to scout for the most suitable representation of the subject and create an appropriate cover for the book.

The publishing team has been an ardent support to the editorial, designing and production team. Their endless efforts to recruit the best for this project, has resulted in the accomplishment of this book. They are a veteran in the field of academics and their pool of knowledge is as vast as their experience in printing. Their expertise and guidance has proved useful at every step. Their uncompromising quality standards have made this book an exceptional effort. Their encouragement from time to time has been an inspiration for everyone.

The publisher and the editorial board hope that this book will prove to be a valuable piece of knowledge for students, practitioners and scholars across the globe.

Index